Synthesis Lectures on Engineering, Science, and Technology

The focus of this series is general topics, and applications about, and for, engineers and scientists on a wide array of applications, methods and advances. Most titles cover subjects such as professional development, education, and study skills, as well as basic introductory undergraduate material and other topics appropriate for a broader and less technical audience.

Vyasa Sai · Haesik Kim · Bernard Fong

6G Enabled Healthcare Systems

Theory, Implementation and Applications

 Springer

Vyasa Sai
Intel Corporation
Folsom, CA, USA

Haesik Kim
VTT Technical Research Centre of Finland
Oulu, Finland

Bernard Fong
Bespoke Assistive Care Solutions
Auckland, New Zealand

ISSN 2690-0300   ISSN 2690-0327 (electronic)
Synthesis Lectures on Engineering, Science, and Technology
ISBN 978-3-031-73848-7   ISBN 978-3-031-73849-4 (eBook)
https://doi.org/10.1007/978-3-031-73849-4

This Springer imprint is published by the registered company Springer Nature Switzerland AG
The registered company address is: Gewerbestrasse 11, 6330 Cham, Switzerland

If disposing of this product, please recycle the paper.

Preface

As we are anticipating the precipice of the upcoming wireless mobile communications revolution, we need a game-changing approach. 6G systems will redefine the cellular networks and their applications. The convergence of internet of things (IoT) networks and healthcare systems are set to change the landscape of patient care and medical services. This book takes an in-depth look into the transformative potential of 6G wireless technology, exploring how its low latency, massive connectivity, high speed and reliability can revolutionize the way our healthcare and medical services are supported.

Embarking into the journey of 6G is not merely yet another evolution in mobile communications, this transformation brings a holistic vision for a more connected, efficient, as well as patient-centric healthcare ecosystem. 6G will not just play a game, but rather change the game. In order to support new requirements and services of 6G, a new blood technology as well as multidisciplinary approaches are required. With the rapid growth of digital health services, solutions, telemedicine, and the Internet of Medical Things (IoMT), the need for robust, high-capacity networks has never been more critical. This book is designed to highlight the synergies between 6G technology, artificial intelligence (AI) and smart healthcare applications; featuring a comprehensive overview of their implications for medical practice, research, and patient engagement.

Each of the seven chapters presented in this book addresses key themes ranging from the underlying technologies of 6G to its practical applications in various healthcare applications. We explore case studies that demonstrate how 6G can enhance remote patient monitoring, facilitate real-time health data exchange, and support AI-driven diagnostics. Furthermore, we discuss the design and implementation aspects that arise as we integrate these smart technologies into healthcare systems.

Our authors include leading experts from telecommunications and healthcare, both in the industry and academia, all of whom share a common vision of unleashing the power of 6G communications for the betterment of human health. These insights will guide readers through the complexities of this emerging healthcare landscape, highlighting the opportunities and challenges that lie ahead.

Folsom, USA   Vyasa Sai
Oulu, Finland   Haesik Kim
Auckland, New Zealand   Bernard Fong

Contents

Introduction

1.1 6G Enabled Healthcare System

Cellular networks have long been implemented as an alternative to high-speed optical networks with the design of flexible deployment solutions. Recent advances in wireless communication systems have made them very reliable with low latency such that modern wireless systems are capable of supporting many different types of healthcare and medical services [1]. One of the major challenges in supporting reliable and efficient healthcare services is to minimize the probability of error, while maximizing the data throughput under the constraint of network resources and conditions.

There are numerous advantages of a wireless system over wireline alternatives. First, mobility is readily supported since no cabling is required for network connection. The device can be moved around without any constraint of having all the messy cables attached. Another important attribute is scalability where a given system can be easily expanded to accommodate extended area coverage, increased data throughput, and serving an increased number of subscribers. A high degree of scalability ensures all these can be accomplished without the need of installing more cables. These are beneficial both in terms of time and financial costs.

1.2 An Overview of Cellular Wireless Communication Leading to 6G

The popularity of wireless communication technology has increased spectacularly over the past decade sparked by the cellular telephony market growth. This is followed by the technological advancement in mobile computing device designs and feature enhancements of mobile phones. The first generation of cellular communication appeared around

V. Sai et al., *6G Enabled Healthcare Systems*, Synthesis Lectures on Engineering, Science, and Technology, https://doi.org/10.1007/978-3-031-73849-4_1

1980 with each new generation appeared roughly a decade later. We have now observed some four decades of evolution from the first-generation analog system to the current mainstream 5G through to the development of 6G.

Perhaps the most significant milestone for comprehensive and pervasive healthcare systems has commenced through the transition from 3 to 4G (Karamachoski, 2019). The 4G standard was developed by the Third Generation Partnership Project (3GPP) which begun in November 2004, built upon the Universal Mobile Telephone System (UMTS) and is also known as Wideband Code Division Multiple Access (W-CDMA). One of the key approaches is the use of MIMO (multiple-input and multiple-output) that utilizes smart antenna technology with multiple antennas at both the transmitter and receiver ends enabling spatial processing for enhanced spectral efficiency [2]. One of the main design objectives of 5G is substantially reduced latency, as well as increased availability and network capacity. These are all essential attributes for a reliable network to support real-time healthcare services.

Digital cellular data service commenced with GSM (Global System for Mobile communications) in the 1990s that operate in the frequency band of around 880—960 MHz, the 2.5 Generation system supports data rate of up to 9.6 kbps. This rate was then increased in 2 stages. The next data standard, High Speed Circuit Switched Data (HSCSD), utilizes multiple TDMA to assign time slots per connection to support a maximum data rate of 38.4 kbps by using 4 channels simultaneously. Followed by General Packet Radio Service (GPRS) that reuses existing GSM radio channels to provide up to 8 channels simultaneously, supporting a maximum data rate of 140 kbps. It is a packet oriented service that will more efficiently transport bursty data. Wireless Cable, also known as MMDS (Multichannel Multipoint Distribution Service), that in turn supports the delivery of primitive healthcare data to subscribers. Using the frequency band of 2.6 GHz, it also included channels for Instructional Television Fixed Service (ITFS). Channelization allows interactive data services to be supported by MMDS.

Ever since the invention of cellular communication, the successful rollout of the next generation is determined by the ability of a network operator and the equipment manufacturers to deliver products and services that complies with the applicable standards and that the operator can introduce enhancements smoothly without any interruption to services. Since the consumer mobile device is the primary connection that links a subscriber to the network operator, one driving force in moving subscribers to the next generation would be incentives for upgrading the mobile device. In many developing countries, especially those with vast geographical coverage, there is an insatiable demand for wireless communications as rapid growth prospers rural areas where legacy plain old telephone service (POTS) infrastructure have not been well established.

Moving to the 6G era, far higher capacity that can potentially be 100 times that of 5G and much lower latency that is substantially shorter than one millisecond. Coupled with location awareness, it is possible that operations such as natural disaster recovery can be supported by 6G networks with mobile edge computational capabilities to be built into

6G networks. Such properties make 6G particularly suited for highly-customized health anomaly detection.

1.2.1 Healthcare Systems Evolving from 3 to 6G

The process of moving from 3G all the way to 6G involved many industry sectors from network operators to healthcare service providers and telecommunications equipment as well as medical device manufacturers, collective efforts transform numerous technical specification documents into physical products that provides services to consumers who will voluntarily pay money for their mobile devices and the recurring usage. The demand for mobile broadband data has been increased substantially over the past decade since mobile phone users are no longer limited to using their phones merely for voice communications. The need for a more advanced cellular communication system than 3G that has been widely used for over a decade leads to the development of self-organizing network (SON) when moving into 4G, with the advantages of self-configuration, self-optimization and self-healing for enhanced efficiency and reliability. LTE SON is an approach of cognitive radio aspects in cellular communication development [3]. Information about the power and spectrum allocation of neighboring cells is necessary in order to maintain the femtocell coverage and minimize cell-to-cell interference. A femtocell unit integrates the functions of a typical base station (eNodeB), similar to that of a Wi-Fi access point and also includes all the core network elements. In 3GPP terms, an LTE femtocell is referred to as Home Node B for HSPA (HNB) and Home eNode B for LTE (HeNB). Femtocells must be able to search the radio channel and evaluate which network resources are available in order to avoid cross-layer and co-layer interference. The core network that only handles data in packet form has the advantages of reduced latency and higher throughput and cost efficiency.

The network performance became adequate for supporting healthcare systems and services when HSPA networks became mature [4]. A wide range of consumer healthcare and medical devices can be linked to different IoT-enabled devices to provide services from general health assessment to mission-critical rescue missions. From the 2G General Radio Packet Service (GRPS) era in the 1990s, consumers expect to be able to retrieve data through their mobile phones. Since then, incentives for switching from one generation to the next are primarily driven by a range of feature enhancement services such as more secure mobile commerce, mobile payments and mobile wallet, feature-rich interactive learning, smooth streaming music and video; as well as more pervasive and reliable mobile health services. There are different technological requirements for each of these services ranging from information security to network stability and connectivity compliance.

Average revenue per user (ARPU), calculated from the total revenue divided by the number of subscribers, i.e. the revenue generated by one consumer subscription per month or annum, is a number that a network operator strives to maximize for profitability. To increase this number, the network needs highly monthly caps and download data throughput in order to generate higher data ARPU, gaining network capacity in terms of both the number of users supported and data throughput. At the same time, reduction to the data cost structure while experiencing a vast increase in mobile data traffic with ever-increasing multimedia services, particularly important with unlimited data or flat-rate subscription plans [5]. All these push for a brand new network in situations where existing network is saturated and that scalability incurs hefty upgrades.

1.2.2 The 4G War Competing Standards: LTE Versus Mobile WiMAX

The evolution of 4G mobile communications entailed both LTE (Long Term Evolution) and Mobile WiMAX ((Worldwide Interoperability for Microwave Access)) due to historical reasons. Following the two paths made standardization for healthcare service support particularly challenging. LTE, based on the 3G communication standard, was built upon the original Global System for Mobile (GSM) communication standard developed by the European Telecommunications Standards Institute (ETSI) Technical Specification (TS), followed by the subsequent development of 3GPP TS. Whereas WiMAX is evolved from the IEEE 802.16 standard (please refer to 6.2.1 for more details) originally designed for fixed wireless communications. While both LTE and mobile WiMAX are designed to provide high-speed ubiquitous broadband wireless access and are IP networks based on OFDM (orthogonal frequency-division multiplexing) technology, both LTE and mobile WiMAX use similar modulation schemes hence they have similar spectral efficiency. To the network operator, LTE provides a more natural upgrade for their earlier GSM or HSDPA (High-Speed Downlink Packet Access) networks. As with other IEEE 802.X family of standards, WiMAX is built upon an open standard that involved a large community of engineers during its development stage. Such openness implies WiMAX equipment is standard and that anyone is permitted to manufacture WiMAX-compliant equipment. In contrast, developers determining the LTE standard through the 3GPP are formed by a group of network operators and equipment manufacturers who have been trading and using proprietary equipment without open competition.

The WiMAX standard, on the other hand, is evolved from the Local Multipoint Distribution Service (LMDS) platform. It was originally intended to serve as a point-to-multipoint (PMP) last-mile solution. LMDS initially rolled out with carrier frequencies in the 10–40 GHz range with frequency selection primarily depending on local licensing as well as geographical region. Generally, topical areas with persistent heavy rainfall tend to utilize carriers of lower frequencies [6]. The radio range is limited to 3 km due to rain fade attenuation. With the network architecture shown in Fig. 1.1, it was primarily designed as a fixed broadband wireless access (BWA) network. In this network, there are three major components, switching where connection to the outside world is provided,

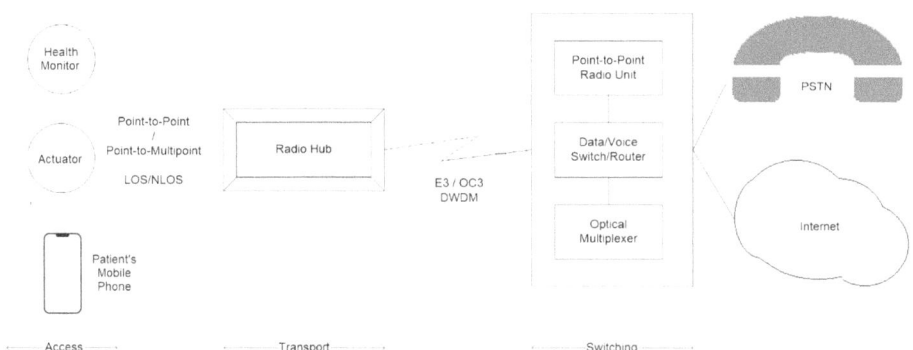

Fig. 1.1 Simple healthcare system architecture

transport where connection to its subscribers are supported, and access where subscribers have their user devices located and are served by the system. Each of these components has a critical role to play in the successful deployment of overall system and they can have a significant impact on the system reliability and performance. One of the key elements of supporting mobility is session continuity for real-time applications including both voice and data communication. So, one fundamental difference between supporting fixed and mobile subscribers is whether connection is provided by one or multiple base stations. This leads to the requirement of "handover" operations.

The licensed frequency band can differ significantly in different countries. For example, in the USA LTE is implemented in the 700 MHz band whereas mobile WiMAX is implemented in the 2 600 MHz band hence LTE has a significant advantage in coverage range over mobile WiMAX as shown in Fig. 1.2. Although 700 MHz offers much wider coverage, MIMO antenna would be much more difficult to implement in mobile phones operating at 700 MHz due to small geometries in relation to the longer wavelength.

1.2.3 Technological Advances in Cellular Networks for Supporting Healthcare Services

Technological advances bring a range of advantages to all parties involved. To the consumers, they expect feature and performance enhancements [7], the network operator drives down the cost of supporting its subscribers in terms of data volume and the number of subscribers through efficient allocation of resources and optimization by network scalability [8]. The equipment manufacturer has the sole objective of selling as many of its products as possible. Attracting as many subscribers as possible benefits both network operator and equipment manufacturer; while driving down the cost per a unit of data (e.g. MB) and provides enhanced coverage is what consumers expect from technological advancements.

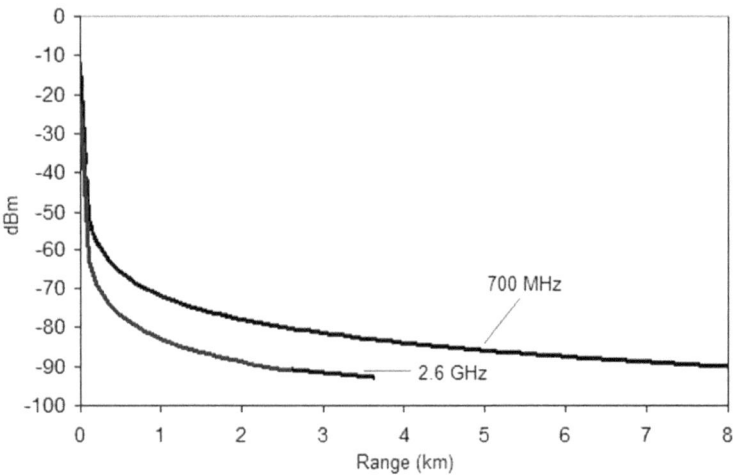

Fig. 1.2 Comparison in coverage between 700 MHz and 2.6 GHz carriers

LTE supports a 50 Mbps upload rate and 100 Mbps download rate for every 20 MHz of spectrum, which is quite adequate for supporting many health monitoring applications like stroke detection [9] and dietary recommendation [10]. The download rate can be boosted to 320 Mbps with a 4×4 MIMO multiple antenna configurations. LTE provides scalable bandwidth from 1.4 MHz to 20 MHz in both the uplink and the downlink.

LTE Advanced (LTE-A) is a mobile communication standard that enhances the capability of the original 4G LTE standard [11]. Perhaps the most important enhancement from LTE is that LTE-A supports a theoretical peak data rate of 1 Gbps when the subscriber moves at a low vehicular speed. Also, LTE-A provides a generic framework for the support of the interworking between different types of legacy cellular networks [12]. LTE-A introduces several features such as Coordinated Multipoint (CoMP) for situations concerning boundaries of multiple cells; Carrier Aggregation for situations where a large chunk of spectrum is split into different sub-channels, for example, five 20 MHz can be aggregated for a total of 100 MHz bandwidth. To combat the effect of multipath using MIMO diversity [13], Multimedia Broadcast and Multicast Service MBMS is configured in such a way that basestations use the same transmission frequency for improved handover.

After competing standards in the earlier 3G (UMTS versus CDMA2000) and 4G (LTE vs. WiMAX), more recent 5G and 6G are developed with one standard. Cross-platform interoperability should be made much easier with a unified standard.

Summary 1.1 The 4G War Competing Standards: LTE versus Mobile WiMAX

1. Substantial enhancements adequate for supporting mission-critical healthcare services commenced from 4G circa 2012 when networks evolved to become mature enough in terms of performance and reliability. Each iteration from 4 to 5G and 6G takes a significant step forward in providing more types of healthcare services to a larger number of patients with more comprehensive coverage (both in terms of healthcare and geographical coverage). A common standard turns out to be vitally important for supporting a diverse range of healthcare services due to issues such as interoperability and compatibility.

1.3   6G-Based Telemedicine Systems Design and Implementation Outline

6G cellular networks with carriers up to sub-THz frequencies are capable of providing wireless connections to many different types of wearable and implantable healthcare and medical devices. These very small devices would require an integrated Terahertz transceiver for 6G wireless communications. Several important factors need to be considered before system implementation. First, wireless communication channels suffer from a number of phenomena such as shadowing and multipath fading especially in an urban environment. The wireless link can be substantially affected by atmospheric conditions such as rain, snow and fog can induce significant signal attenuation to the propagating wave. The effect on attenuation due to relative motion of raindrop and receiver varies significantly that makes 6G telemedicine system's design and implementation particularly challenging.

1.3.1   Built-In Antennas of Healthcare and Medical Devices

Higher carrier frequencies put tremendous demand on requiring a line-of-sight (LOS) link between the base station and device. Maintaining LOS is virtually impossible for mobile devices as the user moves. Near-LOS or LOS requirements lead to significant reduction in cell size due to the use of highly directional antennas. This is particularly problematic for a given base station antenna that is usually pointed directly to one single cell site. Low beam width antennas are often used in mobile devices with appropriate amplification circuitry to compensate for the signal reception properties.

   Path loss in an outdoor operating environment can be greatly affected by factors such as variations in antenna position, diffraction loss due to nearby physical objects, and more significantly, rain attenuation as outlined above. Rain causes signal degradation in the forms of both attenuation and depolarization as the signal propagates towards the receiver. Path loss generally becomes more severe with higher carrier frequencies. One of the challenges in ensuring operational reliability is to maintain a sufficient link margin for the system. Further, shadowing due to movement of physical obstacles that cut through the LOS link can introduce multipath such that multiple copies of the same signal eventually reach the receiving antenna with a slightly different amount of delays. Multipath fading changes substantially over a short distance as a result of the short wavelength of 30mm.

1.3.2 Powering Wearable and Implantable Devices

Electronic devices are becoming smaller and more capable due to advances in data communication, integrated circuit (IC) fabrication and energy storage technologies. The design and implementation of very small healthcare and medical devices very often entail balancing between functional performance and power consumption. There are issues related to battery size and heat dissipation that need to be thoroughly considered [14]. In wearable and implantable devices any heat sink used must be very small as this could severely impact the size and weight of the device. Cooling fan with a DC brushless motor introduces various problems from drawing extra power to generating audible noise and magnetic field hence must be avoided.

Inadequate heat dissipation can cause damage to critical components on the electronic circuit due to overheating. Heat sink is typically made of a thermal conductive metal shaped to maximize surface area that dissipates heat away from the top surface of an IC through thermal convection. The use of metal heat sink, as well as any metal casing, can cause electromagnetic compatibility (EMC) issues [15]. Further, problems related to absorption and skin depth need to be thoroughly addressed [16]. In general, power efficiency must be optimized so that the device draws minimal power from a small power source.

Recent advances in battery chemistry has seen nanostructured silicon-based anodes used to carry a higher amount of charge in lithium ions [17]. Increased in energy density in the battery provides more storage in milli-ampere hour (mAh) so that the device can run longer per charge as well as to shrink the device's size and weight. In wearable and implantable devices where components are closely packed within a very small enclosure, battery cell arrangements, heat transfer materials, as well as cooling channels need to be very carefully designed to prevent cells from overheating that can risk physical expansion [18].

Energy harvesting through mechanical devices can be implemented in lieu a battery so that continuing operation is possible without recharging an internal battery. Figure 1.3 features the back of a mechanical wristwatch that has a mechanical energy harvesting system that can run continuously by winding itself when it is being moved. The basic idea is to power the device through movement [19]. However, radio transmission circuits tend

Fig. 1.3 A mechanical movement found in an automatic winding watch that can potentially be used in powering a small implantable medical device

to draw too much power making this kind of simple mechanism impractical for implementation in implantable medical devices. As an example, an antenna with transmission power in the range of 200—250 mW (23—24 dBm) operating within a small cell that is not much larger than a typical WiFi access point would require a quarter of a Watt for data transmission alone, this excludes the power needed for computation and sensing. Note, incidentally, that the antenna itself does not draw any power given the fact that it is a passive component. Such power is drawn from the radio frequency (RF) signal from the transmission circuit [20].

In terms of the power requirement for data transmission, we looked at the problems associated with path loss in the previous section. Take a carrier frequency of 1 GHz as an example, the path loss L would be measured by:

$$L = 7.3 + \left(log_{10} d \right) \tag{1.1}$$

where d is the distance from the distance between transmitting antenna and the receiver. This roughly translates to a loss close to 4th power of the separation between the transmitter and the receiver [21].

1.3.3   Antenna Power Considerations

Under normal operation, a cellular communication device such as our mobile phone transmit at below 200 mW (23 dBm). The base station should determine the transmission power to a level such that all devices being served by that base station would receive signals of approximately the same strength, i.e. a device that sits right beside the base station should transmit at a minimum power level that would be less than 1 mW. Increase in transmission power will be needed as the device is moved away from the base station and the signal attenuates.

With the popularity of IoT in recent years, devices that are served by base stations are no longer limited to mobile phones as a decade ago. For example, some commercially-available smart watches have direct cellular connection so that direct communication to the outside world is possible without going through the mobile phone. As communication devices and power sources become smaller and more efficient, consumer healthcare and medical devices can be directly connected to the cellular network.

The antenna has a substantial impact on the power efficiency. Generally, the efficiency of an antenna ε_R (a.k.a. radiation efficiency, hence the subscript R is measured by the ratio of power radiated verses to the power supplied to the antenna. The efficiency of most antennas are below 2/3 (-1.5 dB). Some of the power loss is also reflected due to impedance mismatch [22]. Antenna impedance relates the voltage to the current at an antenna's input. An antenna's total efficiency ε_T is the product of the radiation efficiency ε_R and the impedance mismatch loss L_m of the antenna

$$\varepsilon_T = \varepsilon_R \, X \, L_m \tag{1.2}$$

The following factors can minimize antenna loss:

1. Loss due to impedance mismatch;
2. Conduction loss due to finite metal conductivity of the antenna's material;
3. Dielectric loss due to conductivity of dielectric material around the antenna, these include glass, rubber, silicone, plastic and Teflon,

---

Summary 1.2 6G-Based Telemedicine Systems Design and Implementation Outline

1. Energy harvesting through mechanical devices can be implemented for the self-powering of small devices.
2. The antenna has a substantial impact on the power efficiency and several factors can be optimized in antenna design.

References

1. B. Fong, A.C.M. Fong, & C.K. Li, Telemedicine technologies: information technologies in medicine and digital health. John Wiley & Sons. (2020)
2. T. Nakamura et al., Trends in small cell enhancements in LTE advanced. IEEE Commun. Mag. **51**(2), 98–105 (2013)
3. H.Y. Gu, C.Y. Yang, B. Fong, Low-complexity centralized joint power and admission control in cognitive radio networks. IEEE Commun. Lett. **13**(6), 420–422 (2009)
4. B. Fong, K.F. Tsang, C.K. Li, Personalized elderly assistive home care using 3G networks. J. Adv. Inf. Technol. **4**(3), 136–141 (2013)
5. A.A. Toor et al., A robust systematic approach for ensuring optimal telecom service delivery. IEEE Commun. Mag. **58**(8), 49–53 (2020)
6. B. Fong, P.B. Rapajic, G.Y. Hong, A.C.M. Fong, Factors causing uncertainties in outdoor wireless wearable communications. IEEE Pervasive Comput. **2**(2), 16–19 (2003)
7. L. Caviglione, A simple neural framework for bandwidth reservation of VoIP communications in cost-effective devices. IEEE Trans. Consum. Electron. **56**(3), 1252–1257 (2010)
8. B. Fong, N. Ansari, A.C.M. Fong, G.Y. Hong, On the scalability of fixed broadband wireless access network deployment. IEEE Commun. Mag. **42**(9), S12–S18 (2004)
9. T.C. Wu et al., Prehospital utility of rapid stroke evaluation using in-ambulance telemedicine: a pilot feasibility study. Stroke **45**(8), 2342–2347 (2014)
10. A. Yeung, C. & Wu, AI for Elderly Dietary Care: A Cyber-Physical system for nutrition management, IEEE Smart Cities Newsl. **3**(8), (2021)
11. S. Parkvall, et al., LTE-advanced-evolving LTE towards IMT-advanced. *In 2008 IEEE 68th Vehicular Technology Conference.* IEEE (2008), pp. 1–5
12. H. Yokota, A. Idoue, Emerging standards for mobility management in Next-Generation All-IP Networks. *In Invited Paper, Third International Conference on Mobile Computing and Ubiquitous Networking* (ICMU 2006).
13. J. Zhang, N. Ansari, On OFDMA resource allocation and wavelength assignment in OFDMA-based WDM radio-over-fiber picocellular networks. IEEE J. Sel. Areas Commun. **29**(6), 1273–1283 (2011)
14. H.Y. Tung et al., The design of dual radio ZigBee homecare gateway for remote patient monitoring. IEEE Trans. Consum. Electron. **59**(4), 756–764 (2013)
15. W. D. Kimmel, & D. D. Gerke, Electromagnetic compatibility in medical equipment: a guide for designers and installers. CRC Press, 2018.
16. A.D. Porch et al., Electromagnetic absorption in transparent conducting films. J. Appl. Phys. **95**(9), 4734–4737 (2004)
17. P. Li, H. Kim, S.T. Myung, Y.K. Sun, Diverting exploration of silicon anode into practical way: a review focused on silicon-graphite composite for lithium ion batteries. Energy Storage Materials **35**, 550–576 (2021)
18. C.J. Wang et al., Mathematical model for thermal behavior of lithium-ion battery pack under overheating. Appl. Therm. Eng. **191**, 116894 (2021)
19. Z. Lv, L. Qiao, K. Cai, Q. Wang, Big data analysis technology for electric vehicle networks in smart cities. IEEE Trans. Intell. Transp. Syst. **22**(3), 1807–1816 (2020)

20. K. W. Yang, et al., Improving crowd density estimation by fusing aerial images and radio signals. ACM Trans. Multimed. Comput., Commun., Appl., **18**(3), pp. 1–23
21. G. Yuan, X. Zhang, W. Wang, Y. Yang, Carrier aggregation for LTE-advanced mobile communication systems. IEEE Commun. Mag. **48**(2), 88–93 (2010)
22. L. Laughlin, M. ABeach, K. A. Morris, J. L. Haine, Electrical balance duplexing for small form factor realization of in-band full duplex. IEEE Commun. Mag., **53**(5), pp. 102–110, (2015)

Mobile devices have become a part of our lives. We use them in our daily life and cannot even think of how we live without the mobile phones. Mobile devices significantly influence our lifestyle. In particular, people rely on smartphones and manage their daily routine after smartphones are invented. The key role of mobile devices enables us to bridge between humans and exchange information in a variety of ways such as voice and video phone call, email, social media, and so on. In addition, smartphones are multi-functional devices that support scheduling, calculator, multimedia editor, GPS navigator, music player, documentation editor, file storage, and so on. In the 5G era, mobile phones have become even more important devices. Despite these usefulness, there are negative influences of mobile devices on our lives. One of them is the disconnect in our real life. People prefer to interact in cyberspace via smartphones. We can easily observe people using a smartphone in public spaces. They do not have any conversation or eye contact with people. They do not talk to each other, but look at their mobile phone. Ironically, people are more widely connected via mobile devices, but they feel isolated from their family and friends. It is one of the main drawbacks of mobile devices. Another drawback is addiction for people. In particular, children play games and watch videos frequently. Excessive use of mobile devices may cause problems in brain development, eyesight, sleep, nervous system, etc. Recently, a new term Nomophobia (No mobile phobia) has been coined. It means that people feel fear or anxiety about having no mobile devices. Many people suffer from nomophobia and have depression, isolation, or physical problems. Despite these drawbacks, we cannot deny many advantages and big impacts of mobile devices. Cellular systems have continuously evolved from 1 to 5G and have have affected our lives more positively. We need to compensate for the concerns and design

© The Author(s), under exclusive license to Springer Nature Switzerland AG 2025
V. Sai et al., *6G Enabled Healthcare Systems*, Synthesis Lectures on Engineering,
Science, and Technology, https://doi.org/10.1007/978-3-031-73849-4_2

6G systems that are more user-friendly and environment-friendly. In this chapter, we discuss 6G system vision, challenges, requirements, network architectures, and healthcare use cases.

2.1   6G Vision, Requirements, and Timeline

Cellular networks have evolved from the first generation (1G) to the sixth generation (6G). In each generation, new techniques and services are adopted. The 1G is an analogue wireless system based on cellular concept and includes many new wireless techniques such as frequency reuse and so on. It allows users to have mobile voice services. In the 2G, the cellular system becomes more popular and supports voice and short message service (SMS) to mobile users. The problems of 1G such as low system capacity and high-cost handsets were solved in the 2G system by digital TDMA/CDMA and voice compression techniques. From 3G, mobile users require data services. Based on W-CDMA, 3G systems provide mobile users with much higher data services. However, the data services of the 3G systems were not efficient than broadband data services such as ADSL, cable model, or WiFi systems. In 4G systems, improving data services was a key part of 4G networks. The fundamental shift to data services was made by adopting all IP core networks. In the 5G, we achieve another fundamental shift to multiple services and vertical applications. In order to support vertical applications and satisfy the requirements, the 5G networks enhance not only throughput but also latency, connection density, and energy efficiency. The key applications of 5G systems are classified into enhanced mobile broadband communication (eMBB), ultra-reliable and low-latency communication (URLLC), and massive machine-type communication (mMTC). 6G is the sixth generation of cellular systems, which will be the successor of 5G systems and will include new features such as wider frequency bands, higher security, distributed system, open architecture, artificial intelligence, cloud-based networks, and so on. The 5G standards as 3GPPP release 16 and 17 are frozen. The specification of 5G systems has been complete. The specification development of 6G systems is now under discussion in the 3GPP standard. Many research groups in the academy and industry are now focusing on new techniques for 6G communications and networks. It is not easy to predict accurate specification, but the standard contributors may make up for the weakness of 5G systems. For example, the phase II of 5G systems in the 3GPP standard includes URLLC, but it is not enough to fully operate. Therefore, the functions of URLLC may be one of the main parts in the 6G standard. Significant improvement in terms of latency is required in 6G systems. The key applications of URLLC are highly related to control machines, establish tactile internet, and so on. The network reliability should be enhanced together with the low-latency communications. In particular, automated vehicles require extreme high reliability and ultra low latency for steering the vehicles. In addition, 6G will raise the bar for the operation of new applications and services. It will optimize various use cases by adopting new techniques such as artificial intelligence (AI), digital twin, and so on. In particular, AI techniques have matured and been adopted in many different areas. They significantly improve performance and create new applications. Many research groups are actively investigating how to optimize cellular communication systems and networks using AI techniques. The user interface of mobile devices will be improved in the 6G era. From 3G, the touch-screen of mobile devices becomes a key user interface to control the device and make mobile users enjoy the mobile service. In the 6G, it might be replaced by motion or voice control. Mobile devices might be embedded in glasses, clothing, watches, and even skin patches. The 6G system will require new spectrum bands. Each generation of cellular system requires a new spectrum to increase the data rate and accommodate new applications and services. As the 5G systems divide the spectrum into low bands, mid-bands, and high bands, the 6G spectrums will be classified into low bands (below 1GHz) for a low data service like IoT, mid bands (between 1 and 24GHz) for mobile users, and high bands (between 24GHz and 3THz) for high data rate services. The service based network architecture was introduced in the 5G. This approach enables the mobile network to extend the RAN and be deployed in mixed private and public cloud networks in a distributed way. It allows us to have efficient network orchestration solutions, reduce operational cost, and adopt new use cases flexibly.

We expect that 6G systems will take all the capabilities of 5G systems and provide us with a higher level of coverage and performance. Due to IoT systems, massive connectivity is expected and a more complicated network topology will be considered. Data traffic increases exponentially every year. New use cases such as automated vehicles and tactile internet require ultra-low latency communications. Thus, the 6G network should support much higher connection density, higher spectral efficiency, higher energy efficiency, and lower latency than the 5G systems. We can easily imagine how much connection density will increase in 6G systems. For example, the mobile phone was the main device for connecting to radio access networks from 1 to 4G. However, 5G and 6G will expect to connect to many more devices and machines such as mobile phones, tablets, smart watches, smart tags, smart cameras, smart appliances, and so on. It will definitely increase the connection density significantly in 6G systems. Additionally, the resolution of the video increases from HD to 4K. In 6G, holographical video transmission is expected. The data rate of 6G systems should also be improved. Likewise, other network KPIs should be significantly improved. Table 2.1 summarizes the KPI comparison of 5G and 6G.

Adopting AI and machine learning (ML) techniques in 6G is one of the key features compared to the previous generation. The AI techniques will be discussed in Chap. 3. In 6G system, we expect to have a new type of applications such as tactile Internet, haptic Internet, digital twin and so on. They require a new type of system structure and interface between humans and sensors. Massive and ubiquitous sensors of the 6G IoT system will affect our activity, such as sport activity, remote surgery, interventional support in emergency situation, and so on. 5G IoT systems focus on connectivity between cellular networks and IoT devices. 6G IoT systems will expand the IoT systems to ecosystem

Table 2.1 KPI Comparison of 5G and 6G

KPI	5G	6G
Data rate	1–10 Gbps	100Gbps–1 Tbps
Coverage scenario	2D coverage	3D coverage including aerial and space extension
User experienced data rate	100 Mbps	1 Gbps
Peak spectral efficiency	30 b/s/Hz	60 b/s/Hz
The user experienced spectral efficiency	0.3 b/s/Hz	3 b/s/Hz
End-to-End latency	1–5 ms	Less than 1 ms
Reliability	99.9%	99.999%
Block Error Rate (BLER)	10^{-5}	10^{-9}
Connection density	10^6 devices/Km^2	10^7 devices/Km^2
Frequency mid and high bands	3–300 GHz	3GHz–1THz
Position accuracy	1 m	0.1 m
Mobility	500 km/h	1000 km/h
Maximum bandwidth	1 GHz	100 GHz
Satellite and UAV integration	Partially	Fully
Receiver sensitivity	Approx. -120dBm	Approx. -130dBm
Power consumption for IoT	Up to 10 years battery life	Up to 500 years battery life (Zero energy device)

about collaboration between humans and sensors. In the ecosystem, computing power, level of connectivity, and resource allocation will be tailored according to specific use cases. For example, in order to support remote surgery as tactile and haptic Internet, 6G networks should meet special KPI requirements such as high security, ultra-low latency, high reliability, and high throughput. Digital twins also require special connectivity links, such as an always-on feedback loop between real objects and virtual objects. In 6G systems, 5G eMBB will be improved to satisfy new requirements, such as lower latency and higher connection density. The Ubiquitous MBB (UMBB) application will be developed to satisfy high throughput and higher connection density simultaneously. The key use cases of UMBB can be smart city or home with universal high-throughput connectivity with things. Ultra-low-latency broadband communications (ULLBC) will be developed for MBB applications supporting both high throughput and lower latency requirements. eHealth with holographical video transmission and conference will be one use case of ULLBC. Combining URLLC and mMTC is massive ultra-reliable low latency communication (mURLLC). It will be useful for use cases requiring a large number of sensors and time sensitive networks. The key use case of mURLLC can be smart factory that requires industrial sensors and time sensitive control. The new applications (UMBB, ULLBC and

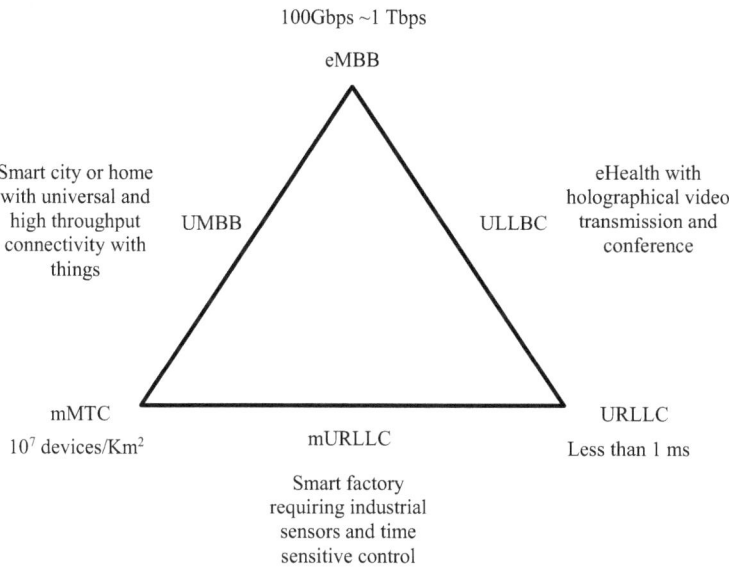

Figure 2.1 6G applications.

Fig. 2.1   6G applications

mURLLC) of 6G systems enable 6G systems to have a much more diverse use cases than 5G use cases. Figure 2.1 illustrates 6G applications. Table 2.2 summarizes the key features of 5G and 6G.

Typically, one generation of cellular systems is 10 years. The 6G systems will gradually evolve for 10 years, including new features. In ITU and 3GPP, requirements, use cases, killer applications, and business models of 6G systems are actively discussed. Figure 2.2 illustrates the timeline of 6G developments.

Summary 2.1 6G Vision, Requirements, and Timeline

1. The key applications of 5G systems are classified into Enhanced Mobile broadband communication (eMBB), Ultra-reliable and low latency communication (URLLC), and Massive machine type communication (mMTC).
2. 6G will set the bar higher to operate new applications and services. It will optimize various use cases by adopting new techniques such as artificial intelligence (AI), digital twin and so on.

Table 2.2 Key features of 5G and 6G

	5G	6G
Key applications	eMBB, URLLC, mMTC	eMBB, URLLC, mMTC, ubiquitous MBB (UMBB), Ultra low latency broadband communications (ULLBC), massive ultra-reliable low latency communication (mURLLC)
Key services	UHD video streaming, V2X, IoT, Smart city, eHealth, and so on	5G key services + Level 4 automated driving, holographic video streaming, tactile internet, etc
Key network features	Slicing, Virtualization, Cloudization	Slicing, Virtualization, Cloudization, Intelligence, Integration
Key technologies	mmWave, Massive MIMO, LDPC and Polar Codes, Flexible Frame Structure, Device-to-device communication, non-orthogonal multiple access (NOMA), Full-duplex, Vehicle-to-everything (V2X), Cloud networks, Edge Computing, NFV and Network Slicing,	5G key technologies + THz Communications, Artificial intelligence, Blockchain-Based Spectrum Sharing, Orbital angular momentum (OAM) multiplexing, Holographic beamforming, (HBF), Large intelligent surfaces (LIS), Visible light communications (VLC),
Key resources	Radio resources (Frequency, time and space), computation	Radio resources, computation, sensing

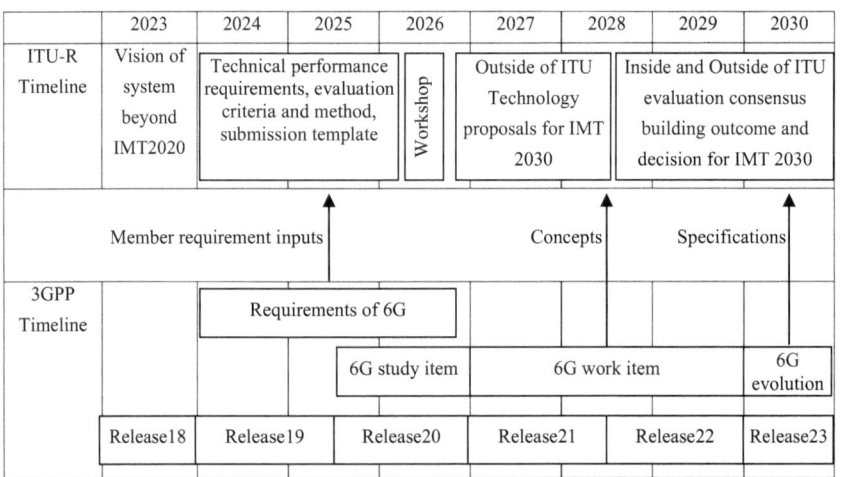

Fig. 2.2 Timeline of 6G developments

3. The service based network architecture was introduced in the 5G. This approach enables the mobile network to extend the RAN and be deployed in mixed private and public cloud networks in a distributed way.
4. In 6G, ubiquitous MBB (UMBB) application will be developed for satisfying both high throughput and higher connection density simultaneously. Ultra low latency broadband communications (ULLBC) will be developed for MBB applications supporting both high throughput and lower latency requirement. Combining URLLC and mMTC is massive ultra-reliable low latency communication (mURLLC).

2.2   6G Spectrum

Since 6G systems will accommodate new applications and services and have high requirements to implement connectivity solution with high demands such as ultra high speed communication, huge volume of data and wide coverage, contiguous and ultra-wide bandwidth should be allocated in 6G spectrum. In higher frequency, we can easily have larger bandwidth but the coverage will be lower because of greater path attenuation of the wave propagation. Thus, the spectrum in high frequency is used for specific indoor use cases requiring extremely high data rates. The spectrum in low frequency is used for wide area coverage use cases with low data rates. In mid bands, we could have compromise on data rates and coverages. 6G spectrum may work in the wavelength ranges centimetric range (7-20GHz) and sub-THz range (above 95GHz). Some experimental works of sub-THz range between 95GHz and 3THz have been tested in many 6G research groups. Furthermore, the 6G spectrum may include some frequencies below 30GHz, including the 4G and 5G spectrums. Coexistence between 6G spectrum and the previous generation spectra may be required by new techniques such as dynamic spectrum. Figure 2.3 illustrates cellular system spectra.

The THz bands as 6G high band give us huge usable bandwidths and allow us to realize 6G use cases requiring the high capacity. However, there are many research challenges to implement THz systems in the real world because they are based on environmental conditions and physical limitations. Key research challenges are the limited coverage, the limited propagation length, difficulty of signal detection, and high cost of THz devices. The mid band of the 6G spectrum will be used in 7–20GHz for urban outdoor users. It can provide us with the necessary bandwidths for new applications and services and achieve high throughput and mobility using new 6G techniques such as massive MIMO. The low band of 6G spectrum will be sub 1GHz (ex. 400–700MHz). It does not accommodate large bandwidth, but enables signals to propagate much further. Thus, it will be useful for IoT solutions. Standardization work is on its way in ITU-R and 3GPP. The availability of 6G spectrum will be determined by country regulators, ITU conferences, and regional

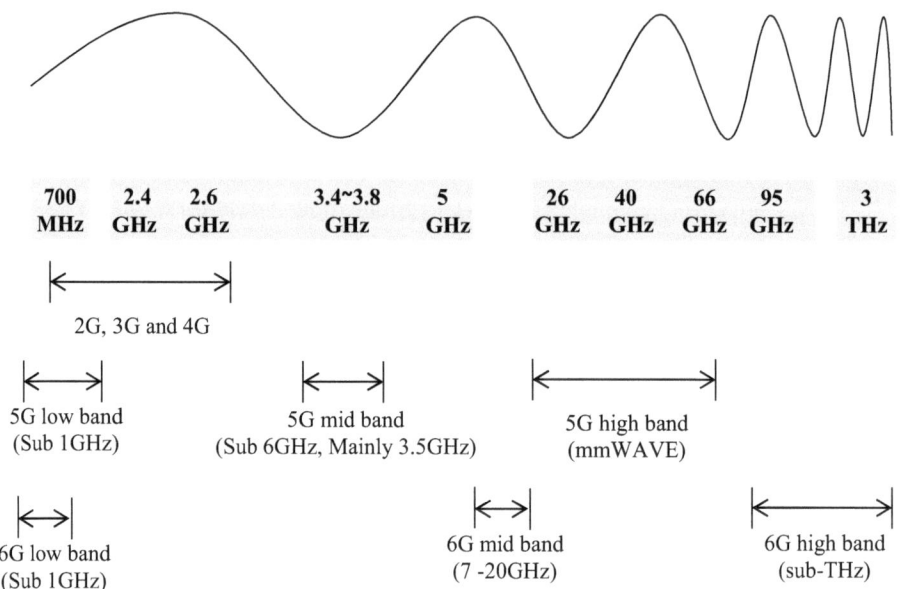

Fig. 2.3 Spectrum of cellular systems

regulators. Harmonization of the 6G frequency will be a key topic. It will be a key starting point for the technological and social success of 6G.

Summary 2.2 6G Spectrum

1. The THz bands as 6G high band give us huge usable bandwidths and allow us to realize 6G use cases requiring the high capacity.
2. The mid band of 6G spectrum will be used in 7–20 GHz for urban outdoor users. It can provide us with the necessary bandwidths for new applications and services and achieve high throughput and mobility using new 6G techniques such as massive MIMO.
3. The low band of 6G spectrum will be sub 1 GHz (ex. 400–700 MHz). It will be useful for IoT solutions.

2.3   6G Network Architecture

Cellular networks (also called mobile networks) have evolved from 1 to 5G. In every generation, new techniques and features have been adopted. The cellular network architectures have been updated. In 1G, the first analogue cellular network is established.

Frequency reuse, cellular concept, FDMA, and licensed spectrum are adopted in the 1G network. Since network capacity was so low in the 1G analog network, the 2G digital cellular network focuses on improving network capacity. The 2G digital circuit-switched networks was established. It provided us with affordable voice services at affordable price. 2G radio access networks (RANs) are composed of the base station subsystem (BSS) and the core network (CN). The 2G BSS contains the base station controller (BSC) and the base transceiver subsystem (BTS). The 2G CN includes the mobile switching center (MSC), the database for mobile user information, the gateway mobile switching center (GMSC), and the interconnection between the 2G network and the PSTN. The 3G cellular network was designed to support a higher data rate than the 2G network. The 3G base station is called eNodeB. It includes new techniques such as WCDMA, congestion control, rate adaptation, accurate power control, congestion control, and so on. In 4G cellular networks, voice and data services are integrated by all IP core networks. Thus, the mobile high speed data services become popular. It is possible to manage 4G cellular networks. The 4G RAN includes a set of eNodeB entities and has control plane and user plane to manage the network efficiently. The 4G CN contains the serving gateway (SGW), the packet data network gateway (PGW), the mobility management entity (MME), and so on. The SGW supports routing and forwarding user data packets and becomes a local mobility anchor for a handover. The PGW allocates IP addresses to mobile devices and supports packet filter and charging. The MME acts in the control plane and supports the signaling and security procedure. The key metrics of 1G to 4G networks are the network capacity to support a high quality voice service and a fast data service. In 5G, the key metrics have been expanded to latency, throughput, energy efficiency, connection density and so on. Thus, it enables the 5G systems to support various use cases and new services. The 5G cellular networks become more intelligent and flexible through virtualization and network openness. Virtualized network components are deployed to create an end-to-end cutting network. In addition, the decomposition and management of network components has been accelerated in 5G. The user plane function (UPF) and the control plane function (CPF) manage user traffic data and the control session, respectively. Due to high requirements and new service, 6G networks should be improved significantly in terms of throughput, latency, energy efficiency, connection density and so on. In order to satisfy the key requirements of the networks and support various services, 6G networks will integrate cellular networks and non-terrestrial networks. Non-terrestrial networks include satellite communications and aerial communications. Satellite systems can be Geostationary equatorial orbit (GEO) that appears stationary at the same point or Low Earth Orbit (LEO) and Medium Earth Orbit (MEO). In particular, many people now pay attention to LEO satellites because a small size and low-cost satellite can be deployed to provide us with wider coverage connectivity. Aerial communications exploit Unmanned Aerial Vehicles (UAVs) that can be deployed instantly and provide us with on-demand connectivity in critical areas. Non-terrestrial networks have key characteristics (global coverage and on-demand infrastructure) that terrestrial networks do not have. Thus, it can compensate for the lack of cellular networks. In particular, non-terrestrial networks will be useful in disaster area or emergency scenario when cellular networks cannot work or cannot be covered. The integration of different networks can be expressed as a network of networks that can interact with each other and find the best connectivity service in an intelligent and self-organizing way. In addition, 6G networks will be flexible, user-centric, and intelligent. The 6G networks will serve a wide range of use cases and applications with high-level KPIs. We will face new challenges. The key challenges of 6G networks are summarized in [1] as follows: (1) Network services will minimize uncertainty and provide mobile users with high reliability. (2) Openness and customization must be adopted to build various ecosystems with vertical industries and support customized services. (3) AI techniques must be adopted to meet the the high requirements of 6G systems. 6G networks should facilitate network intelligence and automation. (4) Integration of the network with satellites is expected to achieve 100% coverage. It will be helpful for 6G network services in a high mountain, artic area, oceans, etc. (5) More efficient spectrum usage is expected. The radio spectrum is a very valuable and expensive resource. Using dynamic spectrum, AI and relevant techniques, we should manage 6G spectrum more efficiently, flexibly, and intelligently. (6) Network security is one of the key techniques in many 6G applications and services. In particular, cellular networks will support autonomous vehicles. If a high level of security could not be guaranteed, we cannot achieve level 5 of autonomous vehicles. In 6G networks, new security techniques such as post-quantum cryptography and quantum key distribution may be adopted to ensure network security. (7) The cell network infrastructure uses high energy and carbon as the network throughput increases. 6G networks must be deployed to achieve the goal of low carbon and energy savings.

2.3.1   Open RAN

Due to high requirements and new challenges, the architecture of the the architecture of the 6G network should be designed in terms of new goals and criteria including AI capabilities, virtualization, flexibility, sustainability, robustness and security, energy efficiency, etc. Innovative cellular networks will be considered in 6G. To achieve a wide range of network targets, open RAN (O-RAN), cloud-based network functions, diverse network deployment and integrated and intelligent networks are actively investigated in many research groups. First, O-RAN enables mobile service providers to easily and efficiently assemble network components according to customer demands. The concept of O-RAN is to open the network protocol, build network interfaces between many different blocks, and allow mobile operators to use mixed network components from different vendors. The key benefits of O-RAN are flexibility, supply chain diversity, low cost of cost of network components, high network performance, interoperability, and so on. We expect to decrease vendor monopolies. Traditional radio access networks rely on network components from big vendors. There is a big market barrier in which small RAN vendors cannot jump in. One of the reasons is interoperability between network components from different vendors. The openness of O-RAN leads to fair competition between big vendors and small vendors. The multi-vendor support of O-RAN will be helpful for optimizing event-based networks. When deploying networks for a big event at concert hall or sports stadium, the flexibility of O-RAN allows an event organizer to have optimized networks and specialized services in a certain period of time. The key features of O-RAN will be open interface, cloudification, and automation. The O-RAN is disaggregated into the Radio Unit (RU) transmitting and receiving radio signals, the Distributed Unit (DU) performing the computation near the RUs, and the Centralized Unit (CU) computing nearby the core network. The CU can be divided into the Control Plane (CP) component and the User Plane (UP) component. In addition, RAN intelligent controller (RIC) is a software component of O-RAN and is responsible for controlling RAN functions. The use of AI or ML is important to manage the intelligent controller of RAN and optimize the network. There are two RICs that manage the network at near-real time and non-real time [2]. The near-real-time RIC is placed at the edge of the network and has control loops with a short time period, such as 10 ms or 1s. It is useful for latency-sensitive application. It is associated with multiple RAN nodes and affects the QoS of thousands of mobile devices [2]. The nonreal-time RIC is part of the service management and orchestration (SMO) framework and manages intelligent RAN operation with larger time scale than 1s. The O-RAN alliance [3] defines multiple interfaces (A1, E2, O1, O2, and so on) such as fronthaul between RU and DU, midhaul between DU and CU, and backhaul to connect between the O-RAN and the core network. They also standardize the virtualization of the O-RAN and extend the evolved Common Public Radio Interface (eCPRI) to connect the RAN RAN node [4]. Figure 2.4 illustrates the open RAN architecture.

The E2 interface is an open interface between the near-real-time RIC and DUs or CUs. It enables the RIC to manage radio resources and control functionalities and procedures of the E2 nodes and also collect metrics from the RAN to the near-real-time RIC. The E2

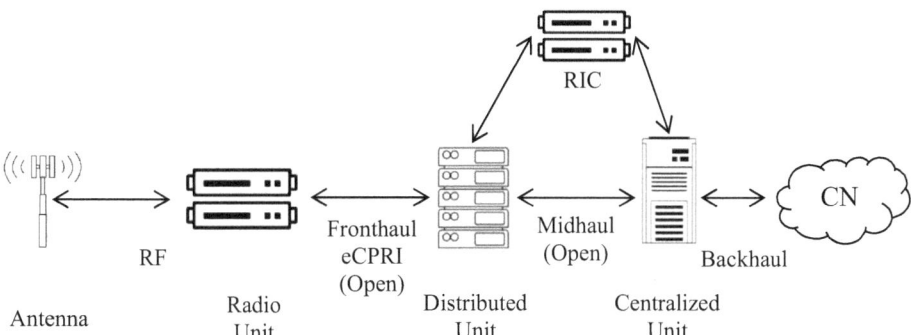

Fig. 2.4 Open RAN architecture

interface runs on top of the stream control transmission protocol (SCTP) over IP. There are two E2 interface protocols, such as the E2 Application Protocol (E2AP) [5] and the E2 Service Model (E2SM) [6]. The E2AP as a basic procedural protocol coordinates how the near-real-time RIC and E2 nodes communicate with each other and provides a basic set of services [5]. The E2AP messages can be embedded with different E2SMs, which implement specific functionalities (namely, the reporting RAN metrics or controlling RAN parameters) [6]. If the E2 connection is complete, the E2AP provides us with services such as E2 report, insert, control, and policy. The E2SM can be used for Key Performance Matrix (KPM), Network Interfaces (NIs), and RAN Control (RC). The O1 interface as an open interface is used for establishing a logical connection between the non-real-time RIC and SMO and ensure the operation and maintenance such as fault, configuration, accounting, performance, and security (FCAPS). The O1 interface can be used to collect data and train the AI model. The A1 interface is used for establishing connection between RICs and supporting policy management, AI management, and data transfer. The AI enrichment information service assists in model training for machine learning and improves RAN performance. In the O-RAN, AI techniques can be accommodated. It offers policies on cell-level optimization by providing a time-varying optimal configuration setting via the O1 interface. The O-RAN Working Group 2 studies the AI/ML workflow and was standardized [7]. The AI/ML workflow has mainly six steps as follows [7]: (1) data collection and processing, (2) training, (3) validation and publication, (4) deployment, (5) execution and inference of AI / ML, and (6) continuous operations. Based on this workflow, mobile operators are able to collect data, train a model, and deploy xApps that are a software component of O-RAN and are used for controlling RAN slicing configuration and optimizing resources and functions of O-RAN according to the network conditions and traffic demand. The AI/ML workflows are summarized in [8]. Figure 2.5 illustrates the AI/ML workflows of O-RAN:

As we can observe in Fig. 2.5, the first step is data collection from O-RAN nodes via A1, O2, and E2 interfaces. The collected data are required to preprocess for training and on-line inference as reshaping the data format. The preprocessed data are used for AI/ML model training. Once the AI/ML model is trained enough, we perform validation process and check whether or not the model is reliable. If the model is valid, they are published on the non-RT RIC. If it is not reliable, AI/ML model is re-trained. If we have a well-trained AI/ML model, they are deployed via the O1 interface. After deploying the AI/ML model on the inference host, we perform inference tasks such as prediction and classification. They can derive network policies at both RICs, take management and control actions via A1, O1, and E2 interfaces. To positively affect network performance, the AI / ML workflow includes continuous operation. As they monitor and verify the inference of AI/ML models, they do not decrease the network performance and refine their functionalities. The AI/ML workflows of O-RAN will be useful if we want to intelligently operate the radio access networks with different requirements and deployment scenarios.

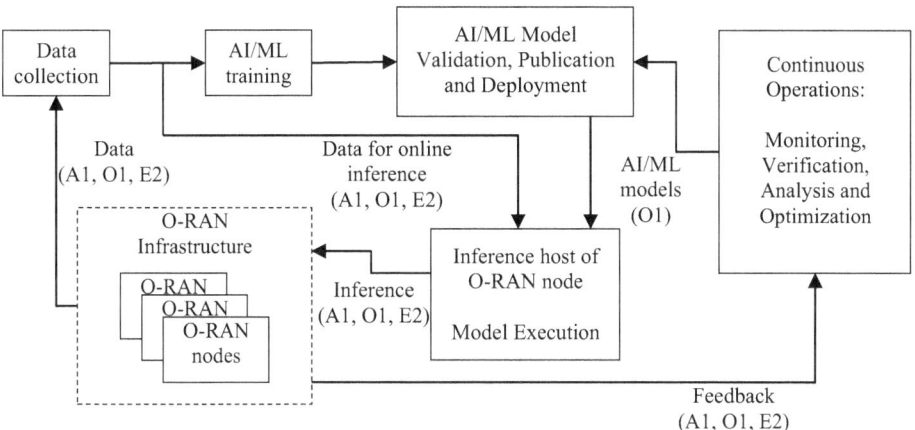

Fig. 2.5 O-RAN AI/ML workflows

2.3.2   User Centric Network

In the era of 6G, different types of networks are deployed in urban areas. Network traffic increases significantly. The number of cells grows exponentially. We face new challenges in deploying networks with higher density. The current network deployment strategies are based on cell-centric networks. This is a traditional way to deploy cellular networks. Multiple mobile devices are connected to one base station. This approach was efficient in providing connectivity to mobile devices in a wide geographical area. However, it typically serves a high data rate for users in cell centers but users in cell boundaries cannot be served properly. Since this large variation in each cell makes the data service quality unstable in the wide geographical area, we need to improve the cellular network performance. In 4G and 5G cellular networks, signal coprocessing was adopted. The basic approach is to perform signal processing at multiple base stations and then transmit to one mobile device. Coordinated multipoint transmission/reception (CoMP), coordinated multipoint with joint transmission (CoMP-JT), and multicell MIMO cooperative networks are based on this approach. In 6G systems, this approach will be expanded to distributed antenna systems or user-centric networks. In user-centric networks, one mobile device can be served by multiple base stations. The key advantages are lower inter-cell interference and higher spectral efficiency. Figure 2.6 illustrates examples of cell-centric and user-centric networks.

As we can observe in Fig. 2.6, the base station in cell centric networks establishes a single connection to mobile devices. The connection between mobile devices and base stations is decided locally. The resource allocation and scheduling is the main task of a single base station. On the other hand, multi-connectivity in user-centric networks is

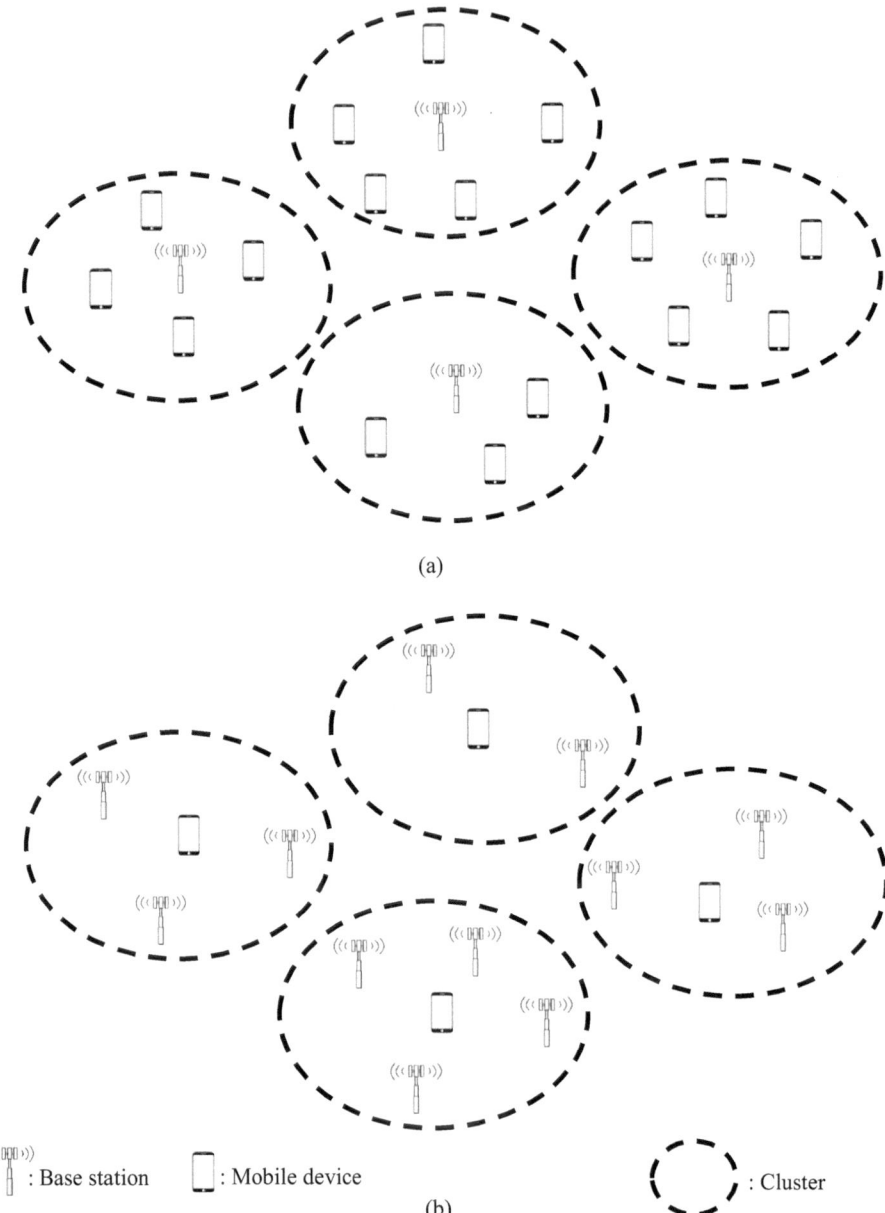

(a)

(b)

Fig. 2.6 Examples of cell centric networks (**a**) and user centric networks (**b**)

provided. Mobile users can switch between available connections. The user centric networks are continuously finding the best available connection and service. Mobile devices influence network decisions by notifying their requirements and conditions to the network. The resource allocation and scheduling are performed in a network centric way. The key challenges to implement user-centric networks are synchronization and reference signals because it is important to find the best cell and acquire frequency and time synchronization. Frequent handover may cause data losses, ping-pong effect, and so on. In particular, since the characteristics of wireless connectivity are varying, key research challenge makes the control plane find the best available connection and handle multiple connectivity. The basic user centric network (or cell free network) architecture is summarized in [9]. A user centric network contains geographically distributed base stations (or access points) that can jointly serve the mobile devices. The base stations are connected to a base station controller (or a central processing unit) via a fronthaul. The multi-connectivity and base station collaboration is managed by a base station controller. Figure 2.7 illustrates an example of user-centric network architecture with single antenna base station, base station controller, and core network.

As we can observe in Fig. 2.7, a base station controller manages the base station via fronthaul and enables them to jointly serve mobile devices in the geographical area. Using

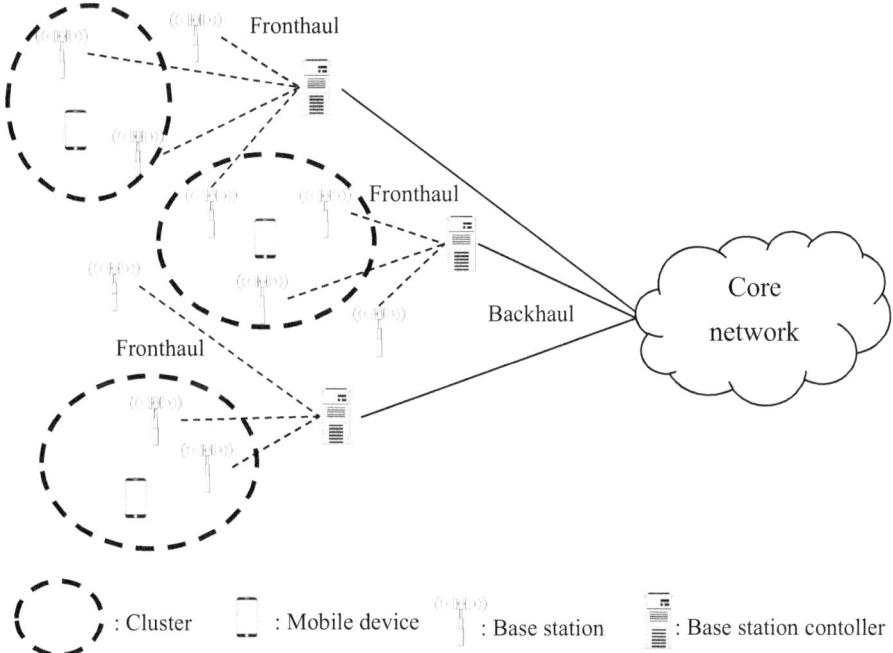

Fig. 2.7 Examples of user centric networks with distributed and cooperative base station and base station controllers

the fronthaul, we can share downlink signals, forward uplink signals, and obtain channel state information. Thus, it enables us to make synchronization between distributed base stations, manage a set of the base station, and have multi-connectivity of mobile stations. We can manage computational capacity more efficiently by sharing computational resources. The base station controller can collect uplink signals and distribute downlink signals to all base stations. Mobile devices can be served by a set of base stations. There are no cell boundaries in terms of mobile devices. The user-centric network is jointly serving mobile devices in the coverage area. If the clusters serving mobile devices are well designed, the performance of user-centric networks will be better than that of cell-centric networks [9]. One of the key design factors is to have standardized interfaces between different types of base stations and access points. Thus, the O-RAN approach will be helpful for designing user-centric networks. Another important design factor is to share the channel state information with the geographically distributed base stations. In the 6G heterogeneous networks, the network configurations and requirements are complex, such as FDD, TDD, low latency, high throughput, and so on. The base station controller should collect them and find the most suitable network configuration to meet the requirements. The massive MIMO concept of distributed systems is well matched with user centric networks. The cell free massive MIMO concept was published in [10]. In this approach, all mobile devices are served by all base stations, and the cell free concept is considered. In distributed manner, user-centric clustering for massive MIMO was published in [11]. In [9], three key benefits of user-centric networks are summarized as follows: (1) It has smaller SNR variations than cell-centric networks with a sparse deployment of base stations and massive MIMO. (2) It is better to manage interference by joint processing of multiple base stations. (3) Coherent transmission increases SNR. It is better to involve base stations with weaker channels in the transmission than to only use the base station with the best channel.

Summary 2.3 6G Network Architecture

1. The key metrics of 1G to 4G networks are the network capacity to support a high quality voice service and a fast data service. In 5G, the key metrics has been expanded to latency, throughput, energy efficiency, connection density and so on. 6G will expand the key metrics. The various network metrics enable the 6G systems to support various use cases and new services.
2. The concept of O-RAN is to open the network protocol, build network interfaces between many different blocks, and allow mobile operators to use mixed network components of different vendors. The key benefits of O-RAN are flexibility, supply chain diversity, low network component cost, high network performance, interoperability, and so on. In O-RAN, The use of AI or ML is important to manage RAN intelligent controller and optimize the network.

3. In the user centric networks, one mobile device can be served by multiple base stations. The key advantages are lower inter-cell interference and higher spectral efficiency. Multi-connectivity in user centric networks is provided. The mobile users are able to switch between available connections. The user centric networks are continuously finding the best available connection and service.

2.3.3   Integrated Networks and Integrated Sensing and Communication

In 6G, we expect seamless and heterogeneous network services with wider coverage. Thus, WiFi, satellite communications, and unmanned aerial vehicle (UAV) communications are considered to integrate with cellular networks. From 4G systems, 3GPP is investigating integration of WiFi and cellular networks. By integrating two different networks, we can obtain many benefits: (1) The mobile operator can offload network traffic to the wifi network and improve efficiency of resource utilization. (2) Mobile users can have a better service by higher data rate, low costs, and wider coverages. (3) Spectral efficiency can be improved by selecting the most suitable bandwidth from two different systems. Due to many advantages, integration with other communication systems is expanded. The integration between cellular networks and satellite communication is considered in 5G systems. The key advantages of satellite communications are wider coverage, flexibility, and adaptability to critical situation. In 6G, the integration will expand further. Integration with UAV communications is now discussed. Typically, UAVs hover at a low altitude and can provide us with connectivity service in the limited area. Key use cases of integration with UAV communications are 360 live broadcasting, aerial relay nodes or backhauls in critical situations, and so on. Figure 2.8 illustrates an example of 6G integrated networks.

Additionally, smart phones and automated vehicles include many sensors that will play an important role in 6G. Especially, when we consider location-ware use case scenarios such as smart city, automated vehicles, healthcare and so on, sensing will be a key function. Basically, sensing collects information from a noisy environment, and communication transmits and receives information via a noisy channel. Sensing data from the environment will be able to support the management of beam and radio resources. The communication of the sensing data will be able to improve the sensing performance. If we unify both functions and co-design mutually like sensing-enabled communication or communication-enabled sensing, we can expect to improve performance, reduce hardware and operational cost, and create new services. Thus, integrated sensing and communication (ISAC) is considered a key feature of 6G radio access networks. The basic concept of ISAC is to combine sensing and communication systems to use radio resources efficiently and to pursue a trade-off between them and mutual performance gains. The key challenges to integrate them can be summarized as follows: (1) Difficulty

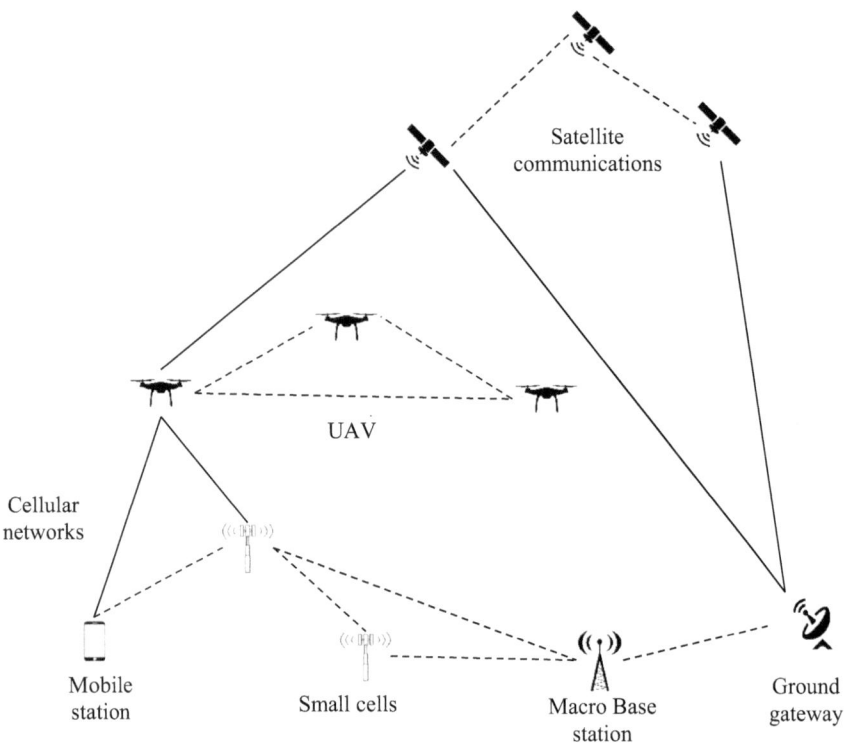

Fig. 2.8 Example of 6G integrated networks

in waveform design, (2) Complicated hardware and resource allocation, (3) Interference mitigation, and (4) Standardization and interface design of heterogeneous systems. In [12], three different approaches to ISAC are presented as joint sensing and communications, sensing-aided communication and communication-aided sensing. Joint sensing and communication has the same hardware and radio resource. Using machine learning techniques, we can design spatial sensing and communication beam patterns, optimize the radio resources, and improve the network security. Sensing-assisted communication improves communication performance using sensory data. Machine learning will play an important role in sensing-aided beam prediction and tracking, sensing-aided blockage prediction, handover among heterogenous networks, and so on. Communication-aided sensing helps sensors to enhance the sensing capabilities and performance of communication signals. Machine learning enables us to optimize communication configuration and design communication-aided and distributed sensing. Automated vehicle scenario can be one use case scenario of ISAC. Figure 2.9 illustrates an example of joint sensing and communications, sensing-aided communication and communication-aided sensing.

Fig. 2.9 Example of joint sensing and communications (**a**), sensing-aided communication (**b**), and communication-aided sensing (**c**)

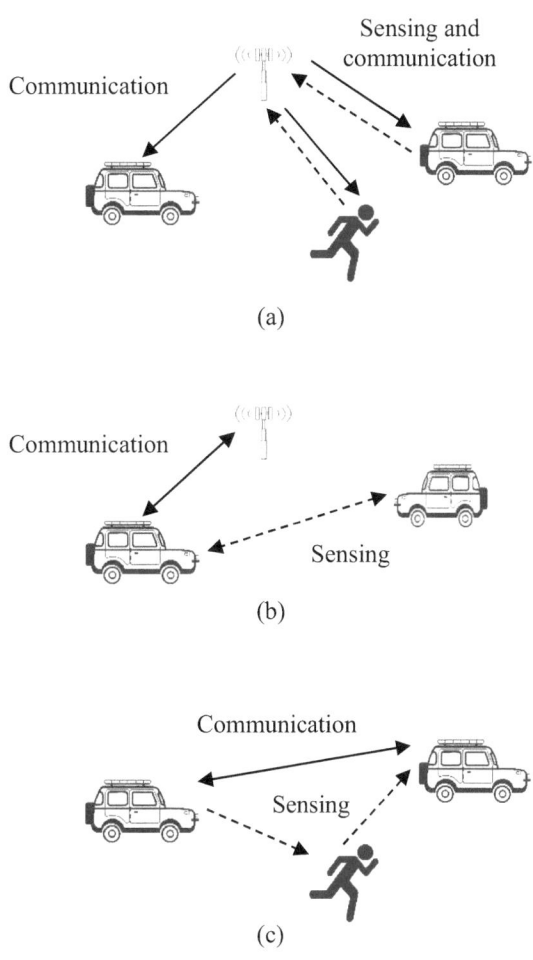

Conventional vehicle detection and communication are separate. There is no intercon-nection between the communication systems and the sensing system in vehicles. They use separate spectrum and hardware. As we can observe in Fig. 2.9, ISAC enables us to combine them and improve driving performance by sharing sensory data and improv-ing communication performance. Although people pay attention to ISAC in 6G systems, combining sensing (such as radar) and communication was considered in 1960 [13]. The combined system as a radar via pulse interval modulation was implemented, where the information was embedded in a set of radar signals. ISAC research and develop-ment was initiated by the Advanced Multifunction Radio Frequency Concept (AMRFC) Program in the 1990s. The AMRFC program is to partition multiple antennas into dif-ferent functional modules and design integrated RF front-ends. The MIMO radar is also a good example to unify communication and sensing. The basic concept of diversity and degrees of freedom from MIMO is used for the MIMO radar theory [14, 15]. In 6G, we expect to further integrate sensing devices and communication systems. There are literally sensing devices everywhere. It is not subsidiary equipment but a basic device in the ICT infrastructure. 6G networks will evolve into a network that can sense the environment and provide us with new services such as human activity recognition, traffic control, smart city, and so on. The key use cases of ISAC are summarized in [16] as (1) Sensing-as-a-Service (Enhanced localization and tracking, area imaging, mobile crowd sensing, passive sensing network, drone monitoring and management), (2) Smart Home and In-Cabin Sensing (Human presence detection, sleep monitoring, heart rate estimation, intruder detection, sensing-aided wireless charging, passenger monitoring, driver attention monitoring), (3) Vehicle to Everything (Vehicle platooning, simultaneous localization and mapping), (4) Smart Manufacturing and Industrial IoT (Employee Localization and Authorization, Manufacture Defect Analysis, Automatic Guided Vehicles, Predictive Maintenance), (5) Remote Sensing and Geoscience (Drone Swarm SAR Imaging, Satellite Imaging and Broadcasting), (6) Environmental Monitoring (Weather Prediction, Pollution, Rain and Insect Monitoring,), (7) Human Computer Interaction (Gesture Recognition, Keystroke Recognition, Head Activity Recognition, Arm Activity Recognition). Using ISAC, cellular systems become a large-scale sensor network for new use cases and services. Wireless 6G wireless systems will be a good infrastructure to implement sensing techniques. The integration between IoT devices and cellular networks can be efficiently implemented by reusing reference signals as a sensing waveform [16].

Summary 2.4 Integrated Networks and Integrated Sensing and Communication

1. In 6G, the integration with UAV communications is now discussed. Typically, UAVs are hovering at a low altitude and can provide us with connectivity service in the limited area. Key use cases of integration with UAV communications are 360 live broadcasting, aerial relay nodes or backhauls at critical situations, and so on.
2. The basic concept of ISAC is to combine sensing and communication systems to use radio resource efficiently and pursue a trade-off between them and mutual performance gains. The key challenges to integrate them can be summarized as follows: (1) Difficulty of waveform design, (2) Complicated hardware and resource allocation, (3) Interference mitigation, and (4) Standardization and interface design of heterogenous systems.

2.3.4 6G Intelligent Network

We can get a sense of the 6G intelligent network through ETSI activities. ETSI Industry Specification Group (ISG) Experiential Networked Intelligence (ENI) investigates intelligent network architecture for network management, considering context-sensitive policies to adjust network service provisioning according to user needs, environmental conditions, and business goals [17, 18]. ETSI ISG ENI develops an automated network with a decision-making process to improve network performance. In [19], the functionalities of the ETIC ISG ENI systems are summarized as shown in Fig. 2.10. They deal with how intelligence is applied to the network and applications to improve network operations, management, and experience of service provisioning [19].

They defined service and network requirements, functional requirements, and non-functional requirements. These requirements are used to evaluate how well AI techniques work in the network to improve network performance. Network intelligence covers (1) the network management including planning, maintenance, and service allocation, (2) the network operation to optimize the configuration, (3) the orchestration of services considering different service level agreement of vertical applications, (4) the network monitoring and prediction of network status, and (5) AI technique implementation. The generic autonomic network architecture (GANA) of ETSI Generic Autonomic Networking Architecture (GANA) [20] is similar to the self-organizing network. The GANA Autonomic Management and Control (AMC) contains an autonomic decision-making element

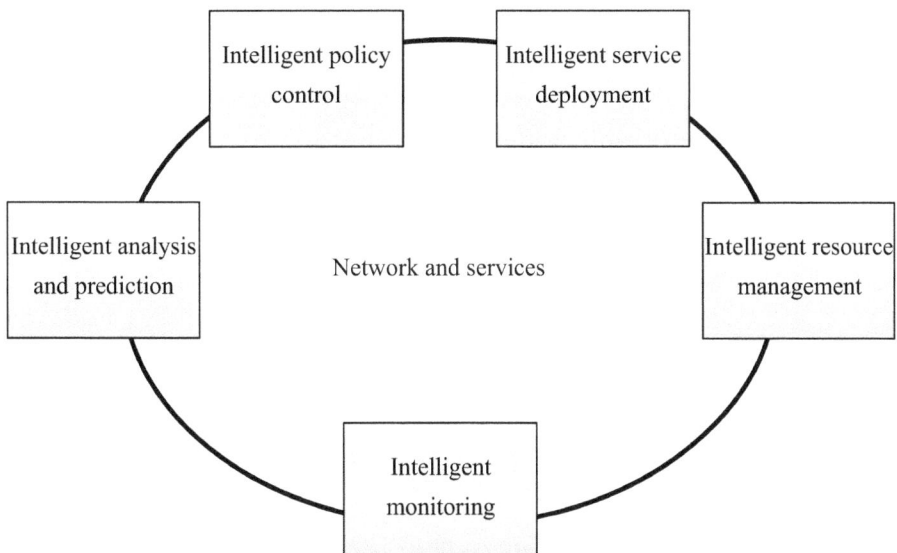

Fig. 2.10 Functionalities of the ETSI ISG ENI system

(DE), that is, a cognitive control loop in centralized or distributed management and control planes [19]. The DE enables us to monitor and manage the network. The role of AI will be critical to realize the intelligence and automation of 6G networks. The key challenges of AI techniques for the 6G intelligent network will have a good training model. Since the network conditions are varying, a a single AI technique could not provide us with a a general solution. Thus, multiple AI models may be used to adopt different network conditions.

Summary 2.5 6G Intelligent Network

1. We can get a sense of 6G intelligent network through ETSI activities. ETSI industry specification group (ISG) Experiential Networked Intelligence (ENI) investigates on intelligent network architecture for network management, considering context-aware policies to adjust network service provisioning according to users' needs, environmental conditions, and business goals [2.17, 2.18].
2. The AI role will be critical to realize intelligence and automation of 6G networks.

2.4   6G Healthcare Use Cases

The World Health Organization (WHO) defines e-health as the transfer of health resources and healthcare by electronic means, including the delivery of health information through the the Internet and telecommunications, improving public health services through education and training of health workers, and finally using electronic commerce and electronic business practices for the management of the management of health systems [21]. 6G networks will be helpful for monitoring, diagnosing, and treating patients anywhere, anywhere, at any time. Artificial intelligence techniques will allow access to 24/7 healthcare service from anywhere, enabling continuous monitoring as well as reliable reliable low-cost diagnostics. We can observe key healthcare use cases of 5G systems from the EU 5G-HEART project [21]. The vertical development of 6G healthcare will be in line with 5G healthcare use cases. In [21–23], 5G healthcare use cases are implemented. All healthcare use cases share a common vision of remote care or 'hospitals without walls.' Table 2.3 summarized the key healthcare vertical of 5G systems.

The 5G-HEART use case scenarios are summarized as follows [21]:

- H1 (H1A 'Educational surgery', H1B 'Remote ultrasound examination', H1C 'Paramedic support', H1D 'Critical health event') – 'Remote interventional support' that explores the use of advanced, rich media communications in the context of remote monitoring, education, and robotics in patient diagnostics and treatment.

Table 2.3 Healthcare use case of the EU 5G-HEART project [21]

Use case H1: Remote Interventional Support	Use case H2: Automatic pill camera anomaly detection	Use case H3: Vital-sign patches with advanced geolocation
The use of advanced, rich media communications for remote monitoring, education, and robotics in patient diagnostics and treatment	Colon wireless capsule endoscopy with automatic polyp detection for early detection of colon cancer with high mortality	Direct-to-Cloud, disposable, vital sign patches to enable continuous monitoring of ambulatory patients, anytime and anywhere
– Educational surgery – Remote ultrasound examination – Paramedic support – Critical health event	– Pill-based endoscopy for early anomaly detection – Remote wireless capsule polyp detection	– Vital-sign patch prototype – Localizable tag

- H2 (H2A "The Pillcam")—"The Pillcam", which aims to test real-time transmission with feedback control of a pill camera (capsule video endoscopy) in order to improve the diagnosis of cancer or precancerous pathology.
- H3 (H3A 'Vital-sign patch prototype', H3B "Localizable tag")—'Vital-sign patches with advanced geolocalization', whose objective is to explore disposable vital-sign patches direct to the cloud to allow continuous monitoring of ambulatory patients, anytime and anywhere.

Figure 2.11 illustrates the network requirement of the healthcare verticals [21].

In the healthcare vertical, the most stringent requirements included UL—DL through-put, latency, location accuracy, mobility, and reliability suggesting the combination of eMBB and URLLC. Energy and power efficiency in relation to coverage turned out to play an important role for use-cases involving battery powered devices suggesting the use of mMTC as well in related scenarios. The 6G healthcare use cases will be similar to the the 5G vertical use cases and equipped with AI-driven technology. 6G healthcare care use cases improve our perception of a healthy life.

High quality video streaming is one of the key features in 5G and 6G networks. In the healthcare sector, it is essential to observe the health conditions or medical images of patients. The educational surgery use case of 5G-HEART is based on high-quality video streaming to observe clinical operations from remote sites. The video streaming platform delivers a real-time video from a surgery room to remote sites such as a classroom or conference room. The basic configuration and scenario of the use case are as follows: A single 360 degree camera is installed in a surgery room or an ambulance and transmits a high-quality video to a remote site via 5G connection with a high data rate. In the remote site, people joins in the session through their own end-user devices such as mobile phone, AR/VR glasses and so on. During surgery or emergency treatment, a doctor can move

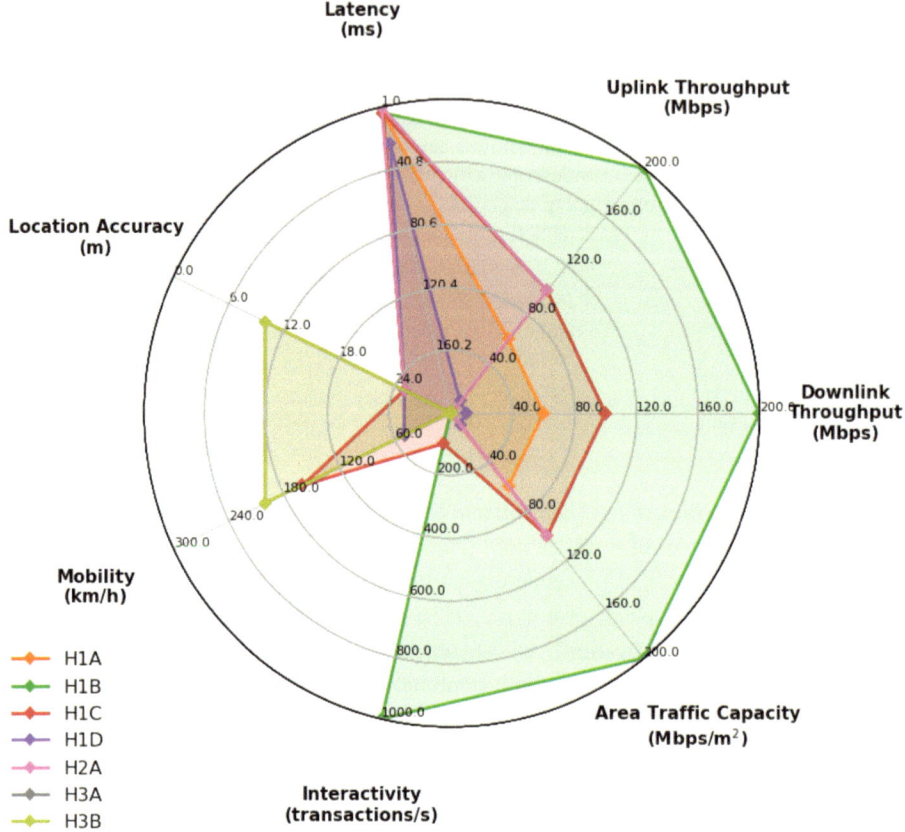

Fig. 2.11 Vertical healthcare network requirements [21]

through a surgery room or an ambulance and people at a remote site can observe the room with 360 degree view. The network configuration between a server side (a surgery room or an ambulance) and a client site (a class room or a conference room) is as follows. The 5G test networks (5GTN) [24] are implemented in multiple cities in Finland. The core network of 5GTN enables us to find evolved packet core (EPC) simulator performing as stand-alone (SA) core, as well as the non-standalone (NSA) cloud mobility manager core (CMM) hosting home subscriber server (HSS) at different locations. A small 5G cell connected to the 5G SA core is deployed for the use case and performs live measurements. Figure 2.12 illustrates the configuration of the 5GTN network for the educational surgery use case.

The KPIs of this use case are throughput and latency. They are measured in the networks as shown in Fig. 2.12. The network throughputs are measured as about 20 and 60

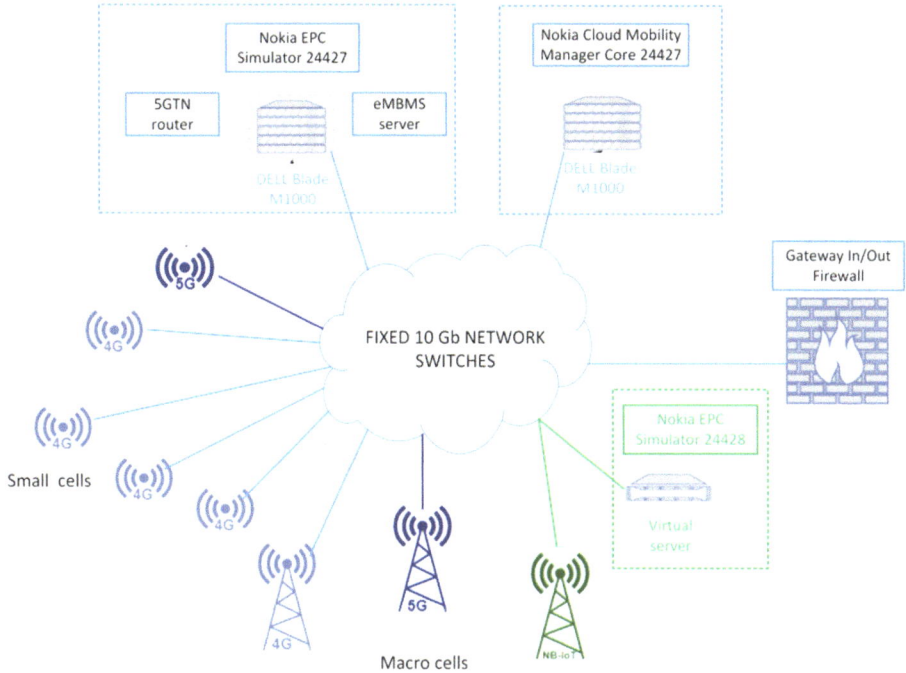

Fig. 2.12 5GTN network for the educational surgery use case [22, 23]

Mbit/s for UL and DL, respectively [23]. In the mobile network, the measured end-to-end latency ranged from 1.74 to 2.28 s and the average was 2.01 s [23]. The end-to-end latency includes streaming video encoding and decoding delay, buffer delay, transmission delay, processing delay, and so on. This is much longer latency than the 5G requirement. This remains a great research challenge to implement time-sensitive use cases.

Summary 2.6 6G Healthcare Use Cases

1. We can observe key healthcare use cases of 5G systems from EU project 5G-HEART [21]. 6G healthcare vertical development will be in-line with 5G healthcare use cases.
2. In the healthcare vertical, the most stringent requirements included UL—DL throughput, latency, location accuracy, mobility and reliability suggesting the combination of eMBB and URLLC. Energy and power efficiency in relation to coverage turned out to play an important role for use-cases involving battery powered devices suggesting the use of mMTC as well in related scenarios.

3. 6G healthcare use cases will be similar to 5G vertical use cases and be equipped with AI driven technology.

References

1. Haesik Kim, *Artificial Intelligence for 6G*, Springer, (2022), ISBN-10: 3030950409
2. O-RAN Working Group 2, "O-RAN Non-RT RIC Architecture 1.0," O-RAN.WG2.Non-RT-RIC-ARCH-TS-v01.00 Technical Specification, (2021)
3. https://www.o-ran.org/
4. O-RAN Working Group 1, "O-RAN Architecture Description 5.00," ORAN, WG1.O-RAN-Architecture-Description-v05.00 Technical Specification, (2021)
5. O-RAN Working Group 3, "O-RAN Near-Real-time RAN Intelligent Controller, E2 Application Protocol 2.00," O-RAN.WG3.E2AP-v02.00 Technical Specification, (2020)
6. O-RAN Working Group 3, "O-RAN Near-Real-time RAN Intelligent Controller E2 Service Model 2.00," ORAN-WG3.E2SM-v02.00 Technical Specification, (2020)
7. O-RAN Working Group 2, "O-RAN AI/ML Workflow Description and Requirements 1.03," O-RAN.WG2.AIML-v01.03 Technical Specification, (2021)
8. M. Polese, L. Bonati, S. D'Oro, S. Basagni, T. Melodia, Understanding O-RAN: architecture, interfaces, algorithms, security, and research challenges. Submitt. Publ. IEEE, arXiv:2202.01032.
9. Özlem Tuğfe Demir, Emil Björnson, Luca Sanguinetti, Foundations of User-Centric Cell-Free Massive MIMO. Found. Trends Signal Process., **14**(3–4), pp 162–472, (2021). arXiv:2108.02541
10. S. Buzzi, C. D'Andrea, Cell-free Massive MIMO:User-centric approach. IEEE Commun. Lett. **6**(6), 706–709 (2017)
11. E. Björnson, L. Sanguinetti, Scalable Cell-free Massive MIMO systems. IEEE Trans. Commun. **68**(7), 4247–4261 (2020)
12. U. Demirhan, A. Alkhateeb, Integrated sensing and communication for 6G: ten key machine learning roles. Submitt. IEEE Signal Process. (2022). arXiv:2208.02157
13. R.M. Mealey, A method for calculating error probabilities in a radar communication system. IEEE Trans. Space Electron. Telemetry **9**(2), 37–42, (1963)
14. J. Li, P. Stoica, MIMO radar with colocated antennas. IEEE Signal Process. Mag. **24**(5), 106–114 (Sept. 2007)
15. A.M. Haimovich, R.S. Blum, L.J. Cimini, MIMO radar with widely separated antennas. IEEE Signal Process. Mag. **25**(1), 116–129 (2008)
16. F. Liu, Y. Cui, C. Masouros, J. Xu, T.X. Han, Y.C. Eldar, S. Buzzi, Integrated sensing and communications: toward Dual-Functional wireless networks for 6G and beyond. IEEE J. Sel. Areas Commun. **40**(6), 1728–1767 (2022). https://doi.org/10.1109/JSAC.2022.3156632
17. ETSI. Improved operator experience through Experiential Networked Intelligence (ENI). (2017), Available online: https://www.etsi.org/images/files/ETSIWhitePapers/etsi_wp22_ENI_FINAL.pdf
18. ETSI. Experiential Networked Intelligence (ENI); ENI requirements. (2020). Available online: https://www.etsi.org/deliver/etsi_gs/ENI/001_099/002/03.01.01_60/gs_ENI002v030101p.pdf

19. Riccardo Bassoli, Frank H.P. Fitzek, Emilio Calvanese Strinati, Why do we need 6G?. ITU J. Futur. Evol. Technol., **2**(2021), 6 Wireless communication systems in beyond 5G era, pp 1–31, Article https://doi.org/10.52953/IROR5894

20. ETSI, GANA—Generic autonomic networking architecture; reference model for autonomic networking, cognitive networking and Self-Management of networks and services. (2016). Available online: https://www.etsi.org/images/files/etsiwhitepapers/etsi_wp16_gana_ed1_201 61011.pdf

21. https://5gheart.org/

22. 5G-HEART deliverables, D3.1: Healthcare Vertical Trial Requirements Definition and Execution, (2019). https://5gheart.org/dissemination/deliverables/

23. 5G-HEART deliverables, D3.2: Initial solution and verification of healthcare use case trials, (2022). https://5gheart.org/dissemination/deliverables/

24. https://5gtnf.fi/

Artificial Intelligence

3

Artificial intelligence (AI), machine learning (ML), deep learning (DL), big data, and neural networks become buzzwords in many different academies and industries. After two AI winters that significantly reduced interest and research funding for AI topics, many research groups have investigated to solve their own problems, adopted AI concepts in many different fields, and achieved a significant performance improvement. Over the past two decades, many research groups have built a stellar record of achievement in the AI area. The gap between fiction films and the real world has been narrowed. AI already became an important part of many research fields such as image processing, voice recognition, communications and networks, VLSI design, and so on. In this chapter, we will introduce AI techniques and workflow and discuss deep learning in detail.

3.1 Introduction to AI

Human intelligence can be defined in many different ways. This is a philosophical question. Depending on different academic areas, they define and research AI from different points of view. However, we can find common key words such as understanding, learning, reasoning, thinking, problem solving, decision-making, and so on. This book deals with science and technology. Thus, we can define intelligence as understanding an environment and learning from an experience, and making a decision to achieve its goals [1]. In the 1950s, the term "artificial intelligence" was coined by John MaCarthy and became an independent research field at the famous Dartmouth conference [2]. There are multiple categories for AI research. One classification is strong AI and weak AI [3]. In [3], strong AI is defined as 'The appropriately programmed computer with the right inputs and outputs would thereby have a mind exactly the same way human beings have a mind'. A more general definition of strong AI is defined as AI can think, act and learn from

© The Author(s), under exclusive license to Springer Nature Switzerland AG 2025
V. Sai et al., *6G Enabled Healthcare Systems*, Synthesis Lectures on Engineering,
Science, and Technology, https://doi.org/10.1007/978-3-031-73849-4_3

experiences like a human. Strong AI is expressed in many science fiction films. Weak AI is generally defined as "Weak AI machines are algorithms programmed by a human and simulate human behavior under meaningful human control. Weak AI machines cannot perform without human intervention" [1]. Software for voice-activated assistance, such as Apple Siri and Samsung Bixby, is weak AI. Weak AI is already all around us. The definition of strong and weak AI represents how much AI machine emulates the capability of a human.

After the AI research field is defined, decline and revival have been repeated so far. Two big declines are called AI winters. Many research groups paid attention to AI development in the 1950s and 1960s. However, we face the first AI winter in the 1970s. They faced big research problems and did not solve them. In 1973, the state of state of state of state of AI research was assessed in the the the the the Lighthill report [4]. They evaluated that AI visions and promises are exaggerated. Many research funding agencies have have reduced the investment of investment of fundamental AI research. In 1980s, AI research groups focused on applied AI research. Special purpose machines were developed. However, the second AI winter appeared in 1987. The performance of the special-purpose machines relies on big data. The storages were very expensive in the 1980s. In addition, the complexity of the machine increased and the control flow was complicated. It failed to deliver the AI goals again. From the 2000s, computing power increases, storage cost decreases, and mobile networks produce big data and exchange data with high speed. We have good environments for AI and have the capability to operate AI algorithms cost efficiently. Deep learning enables us to solve problems that were difficult for other fundamental AI techniques. Thus, many industries and services are now actively adopting AI techniques in their developments and business such as factory automating, market forecast, automated vehicle, face recognition, and so on. Figure 3.1 illustrates the deline and revival of AI research.

AI techniques are not a plug-and-play technique. AI systems require interaction with environments, learn from historical data, improve a training model, and create results. Training from the previous data is a key function. AI systems are not isolated and become part of industrial ecosystems. They will have a huge impact on multiple companies. Many companies already exploit AI techniques and create new products and services such as email filtering, face recognition, text autocorrection, chatbots, automated vehicles, etc. To adopt AI techniques in a business, we need to build AI infrastructures such as data collection and management, AI hardware and software, core AI algorithms, AI applications and use cases, and AI business model.

There are many types of AI algorithms. One key question is how to select one AI algorithm to meet requirements and environments. It is very difficult to find the most suitable algorithm under universal conditions. One specific algorithm enables us to find an optimal solution under the given condition, but the algorithm may not work well on another condition. This can be explained by the no-free lunch theorem [5]. As the name said "there ain't no such thing as a free lunch" [5], they have derived that there is no

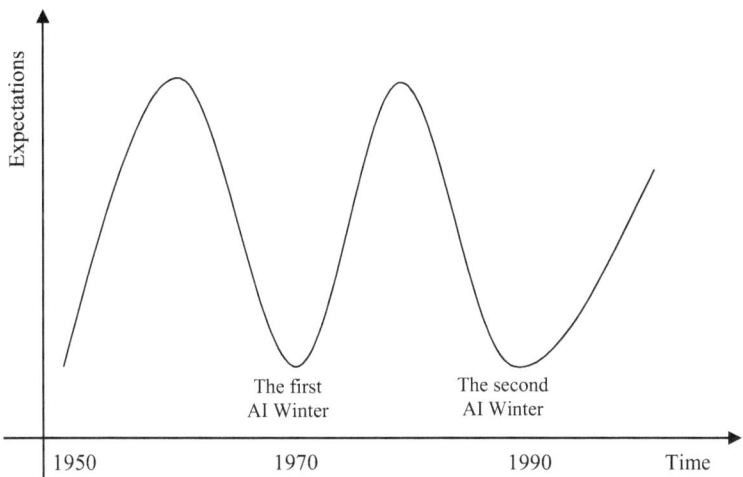

Fig. 3.1 Decline and revival of AI research

universal algorithm to satisfy the universal condition. Thus, the practical way to select one AI algorithm is to understand key features, limitations, and applications of AI algorithms and choose empirically to meet the specific requirements and environments. Figure 3.2 illustrates the classification of AI algorithms.

As we can observe in Fig. 3.2, machine learning is a subset of artificial intelligence techniques. Machine learning can be classified into four types: unsupervised learning, supervised learning, semi-supervised learning, and reinforcement learning. Unsupervised

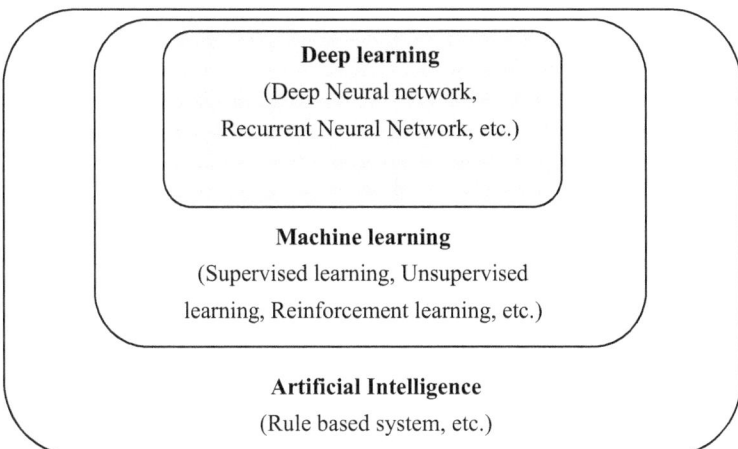

Fig. 3.2 Classification of AI algorithms

learning techniques are used to analyze unlabeled data, observe similarities of data, and discover hidden patterns. They are useful for clustering or reducing dimensionality. Key applications in the real world are to find the target customer group in marketing, to compress data, etc. Supervised learning is used to analyze the labeled data, train a model, and predict unknown data. It is useful for classification and regression. Key applications are image recognition, data forecasting, process optimization, and so on. Semi-supervised learning, as the name says, is a mixed algorithm combining a small amount of labeled data with a large amount of unlabeled data for training. When there are not enough labeled training data, it can enhance accuracy. Key applications are text classification, speech recognition, and so on. Reinforcement learning (RL) is a goal-oriented algorithm to make a sequential decision. It interacts with environments and takes an action to maximize cumulative rewards. RL algorithms are widely used in video games, path-finding, automated vehicles, and so on. Figure 3.3 illustrates three main types of machine learning.

Deep learning uses multi-layered neural networks that are inspired by human brain structure and function. Key applications are face recognition, natural language processing, and so on. We discuss deep learning in the next section.

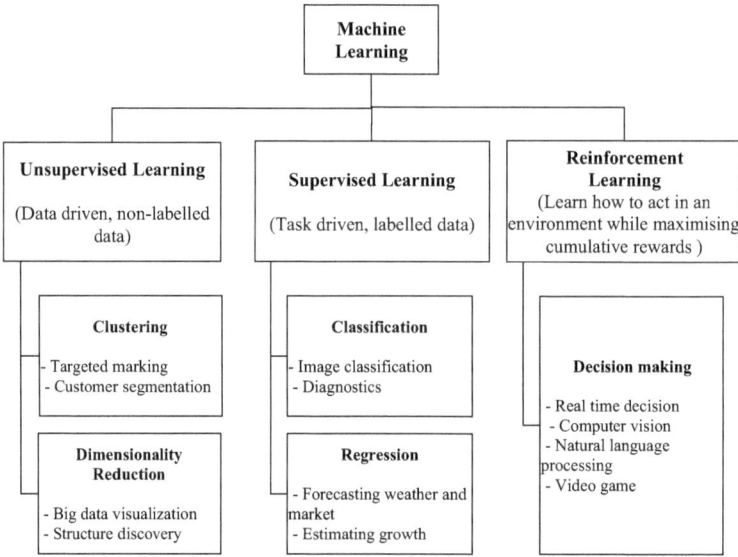

Fig. 3.3 Classification of machine learning

Summary 3.1 Introduction to AI

1. In terms of science and technology, we can define intelligence as understanding an environment and learning from an experience, and making a decision to achieve its goals.
2. General definition of strong AI is defined as AI can to think, act, and learn from experiences like a human. On the other hands, weak AI cannot perform without human intervention. Weak AI is already all around us.
3. There are two AI winters. The first AI winter happens in 1970s and the second AI winter is appeared in 1987. From 2000s, computing power increases, storage cost decreases, and mobile networks produce a big data and exchange data with high speed. We encounter good environments for AI and have the capability to operate AI algorithms cost efficiently.
4. AI techniques are not a plug-and-play technique. AI systems require interaction with environments, learn from historical data, improve a model from training, and create results.
5. Machine learning can be classified into three types: unsupervised learning, supervised learning, and reinforcement learning.
6. Unsupervised learning techniques are used to analyse unlabelled data, observe similarities of data, and discover hidden patterns.
7. Supervised learning is used to analyse the labelled data, train a model, and predict unknown data.
8. Semi-supervised learning, as the name said, is a mixed algorithm combining a small amount of labelled data with a large amount of unlabelled data for training. When there are no enough labelled training data, it can enhance the accuracy.
9. Reinforcement learning (RL) is goal-oriented algorithm to make a sequential decision. It interacts with environments and take an action to maximize cumulative rewards.

3.2   Deep Learning

Deep learning is one of the key AI techniques today. In 1943, the paper 'A Logical Calculus of Immanent Ideas in Nervous Activity' includes the basic concept of neural networks [6]. In 1957, the paper "The perceptron: A probabilistic model for information storage and organization in the brain" contains the deep learning key concept perceptron [7]. The perceptron as a mathematical model of a biological neuron is a basic element of a neural network. In 1986, the paper 'Learning representations by backpropagating errors' includes a backpropagation concept of the neural network [8]. The term "deep learning" was coined in 1986. Deep learning is also called artificial neural networks (ANN). The neural network is composed of nodes and connections that correspond to neurons and synapses in human brains. The structure of the neural network is based on how the human brain works. Thus, the input, output, node, and interconnection of the neural network can correspond to dendrites, axon, cell nucleus, and synapse of the biological neuron, respectively. The node of the neural network receives data, computes them, and passes new data to other nodes via connections. The connections contain weights and biases and affect the calculation of the next nodes. The neural network is based on the neural network with multiple layers. When the neural network has the multiple hidden layers to extract high level features, we call this deep learning where the learning is the process to find optimal weights from data. The basic deep learning process is as follows: (1) Training data, (2) Forward to the neural network for prediction, (3) Finding errors (loss, cost or empirical risks) between true values and predictions, and (4) Updating the weight of connections [1]. Figure 3.4 illustrates an example of a deep learning structure.

As we can observe in Fig. 3.4, the deep learning structure contains an input layer, an output layer, and hidden layers between an input layer and an output layer. The layers contain nodes, and the nodes are connected to each other. Typically, as deep learning structure increases the number of nodes, connections, and layers, the accuracy of the outcomes is improved, but the complexity of computation increases. Therefore, the key design goal is to find an optimal structure between complexity and performance. The design process is carried out repeatedly until a consistent outcome with reasonable results is obtained. Back propagation of deep learning as one of key functions is helpful for improving the performance. It observes an error, updates the weights, and reduces an error. In order to have well-trained deep learning structure, enough number of training

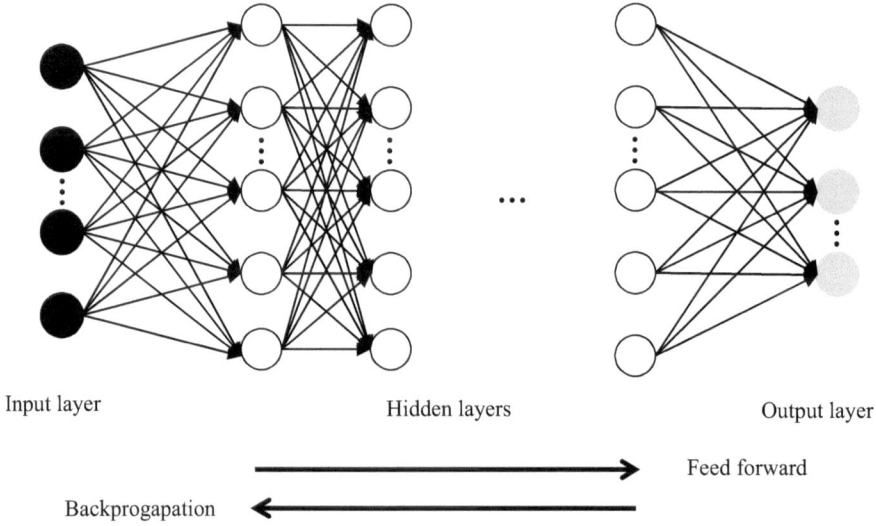

Fig. 3.4 Deep learning with multiple hidden layers and single input and output layer

Table 3.1 Comparison of machine learning and deep learning [1]

	Machine learning	Deep learning
Volume of data and data dependencies	A medium size of data is enough. Typically, thousands of data for optimal point	Big data is required. Typically, millions of data for the optimal point
HW dependencies	Require a less powerful machine than deep learning	Require a powerful machine for a large size of matrix calculation
Output type	Mainly numerical data	Numerical data, image, text and others
Feature extraction	Yes	No
Training time	Low	High
Interpretability	Easy or difficult	Difficult or impossible
Basic approach	Learn through data to solve the problem about classification, prediction, and others	Build neural networks and automatically discover patterns, classification, and others

data is required. Deep learning is a subset of machine learning. The main difference between them is feature extraction. Machine learning has manual feature extraction, and it is not efficient to train a model. When we have a task to classify two different images, conventional machine learning training training should find the the difference between two different images from structured data, mark them manually, and then perform classification. However, deep learning has an end-to-end learning process and does not require manual feature extraction. It does not require structured data and performs feature extraction automatically by different outputs at each neural network layer. Table 3.1 summarizes the comparison of machine learning and deep learning [1].

Now, we look into the deep learning algorithm from perceptron. The perceptron is regarded a single artificial neuron. We can construct a deep learning structure from perceptron. Figure 3.5 illustrates the structure of the perceptron.

As we can observe in Fig. 3.5, the perceptron receives inputs and the inputs are weighted. The bias node is constant. The sum of the weighted inputs becomes the input of a non-linear activation function such as sigmoid function ($f(x) = \frac{1}{1+e^{-x}}$), tanh function ($f(x) = \tanh(x)$), or rectified linear unit function ($f(x) = \begin{cases} 0 \, for \, x < 0 \\ x \, for \, x \geq 0 \end{cases}$). The perceptron enables us to classify the inputs into one of the classes. Assume that the perceptron receives input $\mathbf{x} = [x_1, x_2, \ldots, x_n]$, the output is calculated as follows:

$$\hat{y} = f\left(w_0 + \mathbf{w}^\mathbf{T}\mathbf{x}\right) \tag{3.1}$$

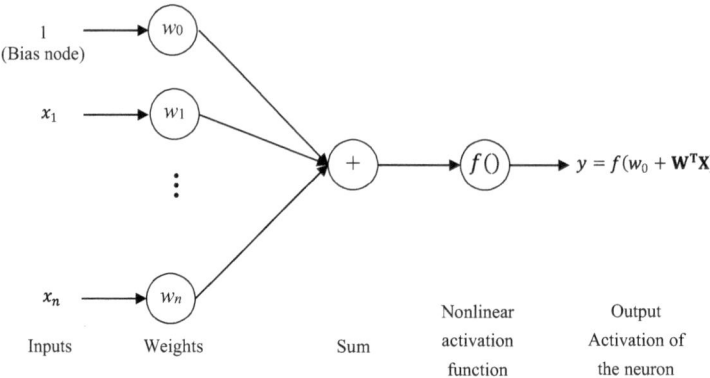

Fig. 3.5 Perceptron structure

where inputs and weights are $\mathbf{x} = [x_1, x_2, \ldots, x_n]$ and $\mathbf{w} = [w_1, w_2, \ldots, w_n]$, respectively. The weights are updated as follows:

$$\mathbf{w}_i \leftarrow \mathbf{w}_i + \Delta \mathbf{w}_i \tag{3.2}$$

where

$$\Delta \mathbf{w}_i = \alpha(d - \hat{y})\mathbf{x}_i \tag{3.3}$$

where α, d and $\hat{y}$ are a learning rate, desired value, and output, respectively. The i is the time index of the data set. The perceptron is used as a linear classifier. The activation function is used to decide the output as a a binary decision. We simplify the perceptron expression to build a neural network and also include multiple output as shown in Fig. 3.6.

As we can observe in Fig. 3.6, it is now similar to the neural network as shown in Fig. 3.4. We can implement logic operations NOT, AND and OR using the perceptron. Figure 3.7 illustrates the characteristics of the logic operation NOT, AND and OR.

What we have to do is to find a proper weight, connection, and threshold. When the perceptron receives two inputs x_1, x_2 and produces output $\hat{y}$, the logic operations can be expressed as shown in Fig. 3.8.

As we can observe in Fig. 3.8, we can simply implement logic operation NOT, AND and OR using a linear classifier like a simple perceptron. However, it is difficult to implement XOR operation from a single perceptron. We should build a more complex neural network to implement more complex algorithms. A multilayer perceptron (MLP) is a fully connected neural network that consists of input layer, output layer, and hidden layers. The hidden layers have specific transformations of input data and help us to have accurate outputs. Figure 3.9 illustrates an MLP with a single hidden layer.

As we can observe in Fig. 3.9, each node of the hidden layer is an output node of a perceptron. The nodes h_j of the hidden layer are represented as follows:

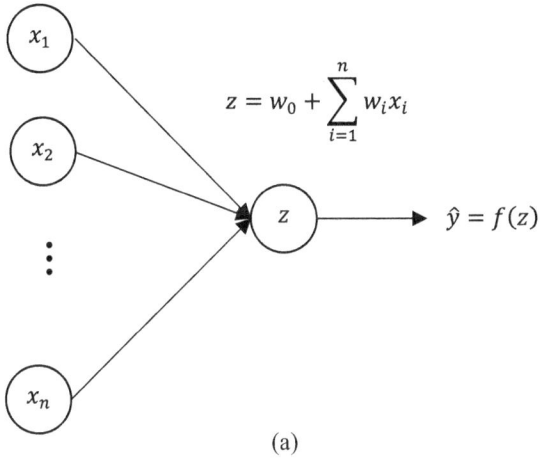

$$z = w_0 + \sum_{i=1}^{n} w_i x_i$$

$$\hat{y} = f(z)$$

(a)

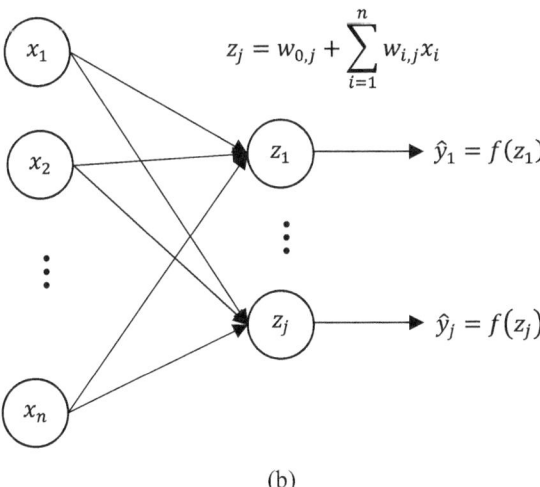

$$z_j = w_{0,j} + \sum_{i=1}^{n} w_{i,j} x_i$$

$$\hat{y}_1 = f(z_1)$$

$$\hat{y}_j = f(z_j)$$

(b)

Fig. 3.6 Simplified perceptron structure (**a**) and perceptron with multiple outputs (**b**)

$$h_j = w_{0,j}^1 + \sum_{i=1}^{n} w_{i,j}^1 x_i \tag{3.4}$$

and the output nodes are connected to the nodes of the hidden layer. They are expressed as follows:

$$\hat{y}_j = f\left(w_{0,j}^2 + \sum_{i=1}^{m} w_{i,j}^2 h_i \right). \tag{3.5}$$

Input 1	Output
0	1
1	0

Input1	Input2	Output
0	0	0
0	1	0
1	0	0
1	1	1

Input1	Input2	Output
0	0	0
0	1	1
1	0	1
1	1	1

(a)   (b)   (c)

Fig. 3.7 Logic operations NOT (**a**), AND (**b**) and OR (**c**)

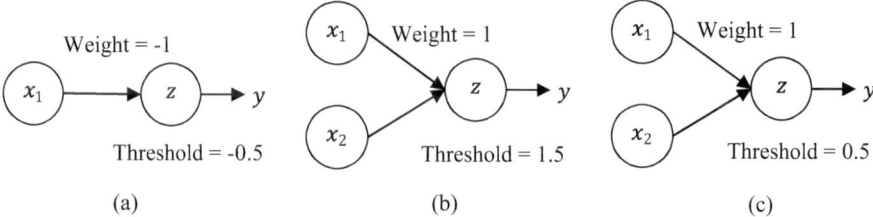

(a)   (b)   (c)

Fig. 3.8 Logic gate implementation using perceptron: NOT (**a**), AND (**b**) and OR (**c**)

Fig. 3.9 MLP with a single hidden layer

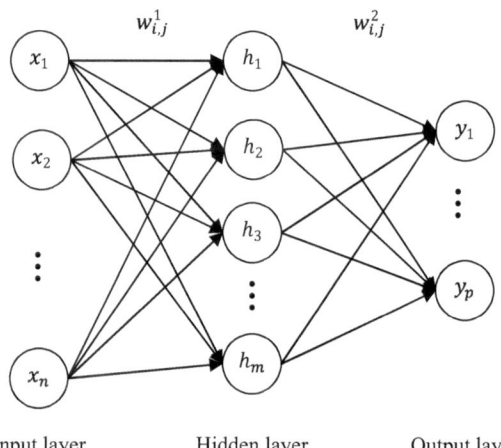

Now, we construct a multilayer feedforward neural network with multiple hidden layers. Figure 3.10 illustrates a multilayer feedforward neural network with l hidden layers.

We assume that a bis node is zero. The first hidden layer nodes can be represented in a matrix form as follows:

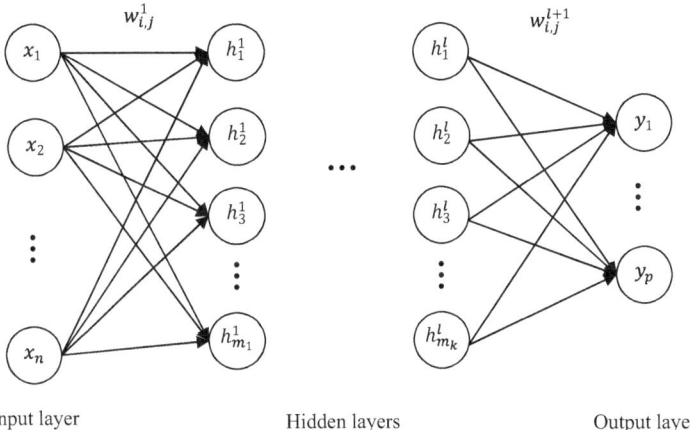

Input layer   Hidden layers   Output layer

Fig. 3.10 Multi-layer feedforward neural network with hidden layer

$$\mathbf{h} = \mathbf{W}^T \mathbf{x}, \tag{3.6}$$

$$\begin{bmatrix} h_1^1 \\ h_2^1 \\ \vdots \\ h_{m_1}^1 \end{bmatrix} = \begin{bmatrix} w_{1,1}^1 & w_{1,2}^1 & \cdots & w_{1,n}^1 \\ w_{2,1}^1 & w_{2,2}^1 & \cdots & w_{2,n}^1 \\ \vdots & \vdots & \ddots & \vdots \\ w_{m_1,1}^1 & w_{m_1,2}^1 & \cdots & w_{m_1,n}^1 \end{bmatrix} \begin{bmatrix} x_1 \\ x_2 \\ \vdots \\ x_n \end{bmatrix}, \tag{3.7}$$

where $\mathbf{h}$, $\mathbf{W}$ and $\mathbf{x}$ are a hidden layer node vector, a weighting vector, and an input vector, respectively. When an MLP has l hidden layers and m_k hidden layer nodes at each layer, the hidden layer nodes are expressed in the matrix form as follows:

$$\mathbf{h} = \mathbf{W}_{m_k}^T \ldots \mathbf{W}_{m_1}^T \mathbf{x}. \tag{3.8}$$

The hidden layer node vectors represent certain features of the neural network. The loss function $J()$ to learn the weights is defined as follows:

$$J\left(\mathbf{W}_{m_{k+1}} \ldots \mathbf{W}_{m_1}\right) = \sum_{i=1}^{N} \|\mathbf{W}_{m_{k+1}}^T \ldots \mathbf{W}_{m_1}^T \mathbf{x}_i - \mathbf{y}_i\|^2. \tag{3.9}$$

We find the optimal weights to minimize the loss function by the backpropagation method and the gradient descent algorithm. As we can observe Fig. 3.10 and Eqs. (3.7) and (3.8), a high-dimensional inputs and a large number of hidden layers result in a high complexity. If a neural network receives N inputs, the number of hidden layer nodes is 2^N. As N increases, the complexity increases exponentially. Thus, if it has many nodes and weights of hidden layer, it can cause an overfitting and slow computation problem.

In deep learning, there are two computations: forward propagation computation and back propagation computation. In forward propagation computation, the inputs are fed in the forward direction. A neural network computes sequentially and passes the output to the next layer until we have the output $\hat{y}$ of the neural network. The forward propagation computation is just mapping from input to output by a simple calculation. There is no learning process. The learning mechanism can be found in back propagation computation. It allows us to find the optimal weights and biases of the hidden layers. The basic idea of backpropagation is to compare the actual values with the desired values and find a loss, cost, or error. The partial derivatives of the loss function are computed in the backward direction. Weights are iteratively updated using gradient descent methods to minimize loss function. When there is an activation function, the loss function (3.9) can be re-written as follows:

$$J\left(\mathbf{W}_{m_{k+1}} \cdots \mathbf{W}_{m_1}\right) = \sum_{i=1}^{N} \left\| f\left(\mathbf{W}_{m_{k+1}}^T \cdots f\left(\mathbf{W}_{m_1}^T \mathbf{x}_i\right)\right) - \mathbf{y}_i \right\|^2. \tag{3.10}$$

The loss function can be updated using the gradient descent method. The update rule for the weights of the first hidden layer can be expressed as follows:

$$\mathbf{W}_1^{t+1} = \mathbf{W}_1^t - \alpha \nabla J(\mathbf{W}_1^t) \tag{3.11}$$

where $\mathbf{W}_1^{t+1}$, $\mathbf{W}_1^t$, and $\nabla J(\mathbf{W}_1^t)$ are new weight, old weight, and derivative of the loss function with respect to weight, respectively. t and α are the parameter for the direction of the gradient and the learning rate, respectively. The back propagation computation starts from the last layer. It propagates the error back to the previous layer and updates the weights of the previous layer. When we have a single hidden layer as shown in Fig. 3.9, the loss function has two weight matrices and can be expressed as follows:

$$J(\mathbf{W}_1, \mathbf{W}_2) = \left\| f\left(\mathbf{W}_2^T f\left(\mathbf{W}_1^T \mathbf{x}\right)\right) - \mathbf{y} \right\|^2 = \|\boldsymbol{\delta}_2 - \mathbf{y}\|^2 \tag{3.12}$$

where

$$\boldsymbol{\delta}_2 = f\left(\mathbf{W}_2^T f\left(\mathbf{W}_1^T \mathbf{x}\right)\right) = f(\boldsymbol{\varepsilon}_2) \tag{3.13}$$

where $\boldsymbol{\varepsilon}_i$ is $\boldsymbol{\varepsilon}_i = \mathbf{W}_i^T \mathbf{h}_{i-1}$. $\mathbf{h}_0$ represents an input layer. $\boldsymbol{\delta}_i$ is $\boldsymbol{\delta}_i = f(\boldsymbol{\varepsilon}_i)$. The partial derivatives of the loss function with respect to $\mathbf{W}_2$ is

$$\frac{\partial J}{\partial \mathbf{W}_2} = \frac{\partial J}{\partial \boldsymbol{\delta}_2} \frac{\partial \boldsymbol{\delta}_2}{\partial \mathbf{W}_2} = \frac{\partial J}{\partial \boldsymbol{\delta}_2} \frac{\partial \boldsymbol{\delta}_2}{\partial \boldsymbol{\varepsilon}_2} \frac{\partial \boldsymbol{\varepsilon}_2}{\partial \mathbf{W}_2}. \tag{3.14}$$

(3.12) can be rewritten as follows:

$$J(\mathbf{W}_1, \mathbf{W}_2) = \left\| f\left(\mathbf{W}_2^T f\left(\mathbf{W}_1^T \mathbf{x}\right)\right) - \mathbf{y} \right\|^2 = \left\| f\left(\mathbf{W}_2^T \boldsymbol{\delta}_1\right) - \mathbf{y} \right\|^2 \tag{3.15}$$

where

$$\delta_1 = f\left(\mathbf{W}_1^T \mathbf{x}\right) = f(\varepsilon_1). \tag{3.16}$$

δ_1 is $\mathbf{h} = f\left(\mathbf{W}_1^T \mathbf{x}\right)$. Likewise, the partial derivatives of the loss function with respect to $\mathbf{W}_1$ can be driven as follows:

$$\frac{\partial J}{\partial \mathbf{W}_1} = \frac{\partial J}{\partial \delta_2} \frac{\partial \delta_2}{\partial \mathbf{W}_1} = \frac{\partial J}{\partial \delta_2} \frac{\partial \delta_2}{\partial \delta_1} \frac{\partial \delta_1}{\partial \mathbf{W}_1} = \frac{\partial J}{\partial \delta_2} \frac{\partial \delta_2}{\partial \delta_1} \frac{\partial \delta_1}{\partial \varepsilon_1} \frac{\partial \varepsilon_1}{\partial \mathbf{W}_1}. \tag{3.17}$$

From (3.14) and (3.17), we can find the sequence of the chain rule. The back propagation can be regarded as the process adjusting weights to reduce the loss function as follows:

$$\Delta \mathbf{W} \propto -\frac{\partial J}{\partial \mathbf{W}}. \tag{3.18}$$

(3.18) can be interpreted as follows:

$$\Delta w \propto -\frac{\partial J}{\partial w} = -\frac{\partial (\text{Loss function})}{\partial (\text{Activation function})} \frac{\partial (\text{Activation function})}{\partial (\text{Network input})} \frac{\partial (\text{Network input})}{\partial (\text{Weight})}. \tag{3.19}$$

The key function of back propagation is the partial derivative of the loss function with respect to weights in the neural network. This equation represents how the loss function affects the change of weights and biases.

There are multiple variants of deep learning [1]. Convolutional neural networks (CNNs) are one of the most popular deep learning models. They have a spatial structure using convolution operation and enable us to have much fewer parameters. CNNs have neurons in 3 dimensions: width, height,, and depth. The layer transforms the 3 dimensional input volumes to the 3 dimensional output volumes of neuron activations. The width and height would be the 2 dimensional image and the depth would be the color channels such as red, green, and blue (RGB). This approach can significantly reduce complexity. A CNN contains four different layers: convolution, activation (Relu), pooling, and fully connected. The convolutional layer is composed of a set of learnable filters that slide over the input. The activation layer allows the neural network to learn harder decision functions and reduce overfitting. The pooling layer is used to down-sample the output of a convolutional layer and gradually reduce the spatial size of the representation. Neurons in the fully connected layer are fully connected to all activations in the previous layer. The common process of CNN can be summarized as follows: input layer Convolution + Relu layer pooling layer ..... Convolution + Relu layer, Pooling layer, Fully connected layer, Softmax layer, Output layer. Figure 3.11 illustrates an example of common CNN architecture.

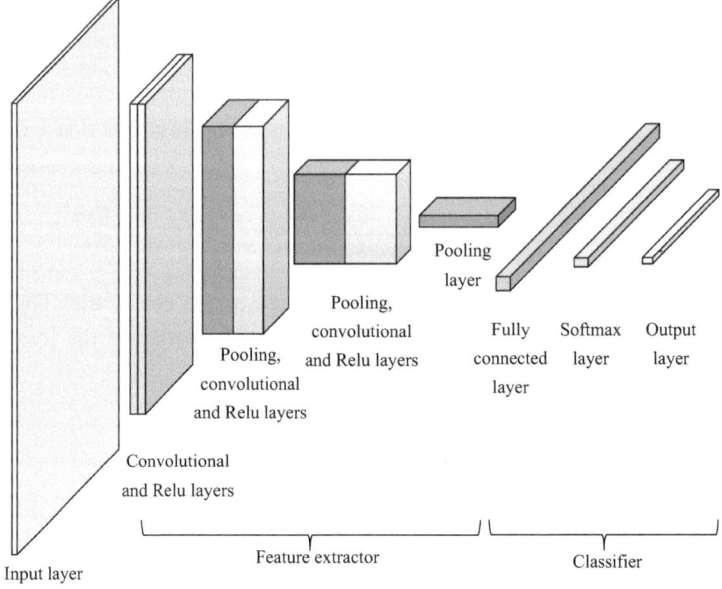

Input layer

Convolutional
and Relu layers

Pooling,
convolutional
and Relu layers

Pooling,
convolutional
and Relu layers

Pooling
layer

Fully
connected
layer

Softmax
layer

Output
layer

Feature extractor

Classifier

Fig. 3.11 CNN architecture [1]

Recurrent neural networks (RNNs) are another popular deep learning model. The basic approach of RNNs is similar to CNNs. They both use parameter sharing. CNNs share the parameters in spatial dimension in pixel image data while RNNs share the parameters in temporal dimension in speech or text data [1]. They exploit the sequential data or time series data to solve common temporal problems such as speech recognition or natural language processing. Training process of RNNs is straightforward. The back propagation method can be applied to the unrolled RNN. We call it backpropagation through time (BPTT). The training data for RNNs are an ordered sequence. The long-short-term memory (LSTM) is a subset of RNNs. It regulates the information flow using some internal gate mechanisms, suppresses vanishing gradients through a gating mechanism, and provides us with better performance. The common process of LSTM can be summarized as follows: it decides whether the information should be kept or forgotten and controls the gradient values at each time step. It allows the neural network to have desired behavior from the error gradient by updating the learning process at each time step. Figure 3.12 illustrates an example of RNN structure. In Fig. 3.12, E, $\mathbf{U}$, $\mathbf{W}$ and $\mathbf{V}$ represent the loss function, the hidden weight matrix input transforming the current input, the hidden-hidden weight matrix transforming the previous hidden state, and the hidden-output weight matrix transforming the new activated hidden state, respectively.

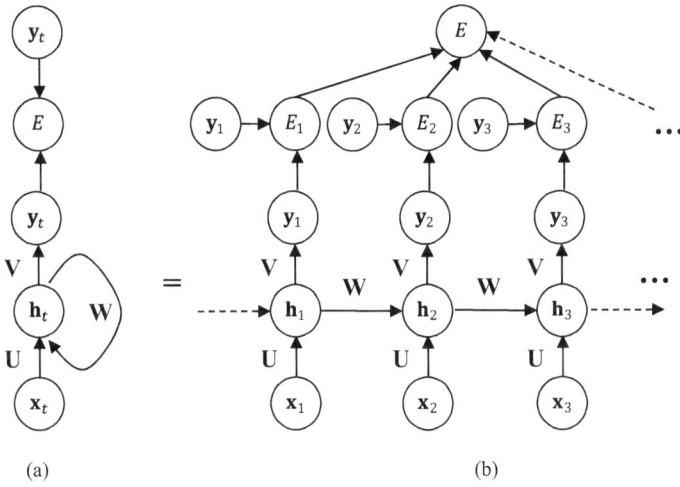

Fig. 3.12 RNN structure (**a**) and unrolled RNN (**b**) including loss [1]

Summary 3.2 Deep Learning

1. The neural network structure is based on how human brains works. Thus, the input, output, node and interconnection of the neural network can correspond to dendrites, axon, cell nucleus, and synapse of the biological neuron, respectively.
2. The basic deep learning process is as follows: (1) Training data, (2) Forward to neural network for prediction, (3) Finding errors (loss, cost, or empirical risks) between true values and predictions, and (4) Updating the weight of connections [1].
3. The key design goal is to find an optimal structure between complexity and performance. The design process is carried out repeatedly until consistent outcome with reasonable results is obtained.
4. The main difference between machine learning and deep learning is feature extraction. Machine learning has manual feature extraction and it is not efficient to train a model. Deep learning has an end-to-end learning process and doesn't require manual feature extraction. It doesn't require structured data and perform feature extraction automatically by different outputs at each neural network layer.
5. In deep learning, there are two computations: forward propagation computation and back propagation computation. The forward propagation computation is just mapping from input to output by simple calculation. There is no learning process. The basic idea of back propagation is to compare the actual values with the desired values and find an loss, cost or error. The partial derivatives of the loss

function are computed in the backward direction. The weights are iteratively updated using gradient descent methods to minimize the loss function.

6. CNNs share the parameters across spatial dimension on pixel image data while RNNs share the parameters across temporal dimension on speech or text data [1].

3.3   AI Workflow and Implementation in 6G

Basically, a workflow is defined as a series of actions that are necessary to achieve a goal. What we want to discuss in this section is how to implement AI systems in cellular communications and networks systems. The overall workflow of AI implementation is to collect data, create a training model to learn a pattern in training data, train a model, validate a trained model, find the pattern in other data, monitor the performance of the model, and manage the model. We can also follow this workflow when implementing AI techniques in cellular systems. The first step data acquisition is a key part of AI implementation. If we do not have enough good quality data, we cannot have a good AI model. The AI model will be as good as the data we collect. Fortunately, 5G and 6G networks are good environments for acquiring big data. We should collect and store enough training data with suitable features and then perform pre-processing such as normalization, filtering, format alignment, and so on. For example, when implementing an AI algorithm for OFDM pilot signal prediction, the collected data could be distorted or significantly interfered significantly due to wireless channel impairments. Data distorted should be ruled out in the training model. The next step in training model creation is selection or development of AI algorithms. In this step, we should determine which algorithm can satisfy the requirements of the overall system and solve the problem efficiently. If we choose one of AI development tools and frameworks such as Google TensorFlow, IBM Watson, Scikit-learn, Amazon Web Service machine learning, Apache MXNet, Microsoft AI Azure, Caffe, Theano, Keras, and so on, we need to select one from the library. It is important to understand the characteristics and limitations of AI algorithms. As we briefly discussed in the previous section, there are some guidelines for choosing an AI algorithm to meet system requirements, but the empirical decision of the AI system developer will be more accurate. For the application of the cellular system, there are some guidelines and examples to select suitable AI algorithms in [1]. For example, AI algorithms will be helpful for the handover between cells. The current handover mechanism is to perform a handover when a mobile user is located on the cell boundary and moves from one cell to another cell. However, if we use the AI algorithm and it predicts future status and makes a decision, we can perform a handover before moving and improve both network capacity and user experience. In this application, the key tasks of AI algorithms are decision making and prediction. Re-enforcement learning and deep learning will be suitable for

Table 3.2   AI algorithm selection for cellular systems tasks [1]

Cellular system tasks / AI tasks	Physical layer	MAC layer	Network and upper layers
Clustering	Signal detection, Symbol mapping	User clustering and association	Critical network establishment, Controlling cluster size in dynamic heterogeneous networks
Classification	Positioning, localisation, MIMO user paring	Dynamic scheduling, Carrier aggregation, Optimal multi-connectivity	Slicing admission, scheduling, Load balancing
Regression	Channel estimation, Equalisation	Adaptive power control	Dynamic slicing management
Prediction	Blind channel estimation	Adaptive power control, Beam prediction, Radio resource allocation	Network monitoring, Orchestration, Traffic prediction, Mobility prediction
Decision making	Error correction coding	Interference mitigation	Handover, Mobility management, Optimal routing path

the handover improvement. Table 3.2 summarizes an example of the selection of the AI algorithm for cellular system tasks.

The next step model training is a process to run the model, adjust its parameter setting, and tunes the model. Depending on the training data we acquire, the AI model is ready to evaluate. The next step of training model validation is the step to test the trained model with different known data. If the performance is good enough, we release the trained model without change. If it is not enough, we tweak the model and improve the performance. Finally, the AI model is ready to be used if the validation step is passed. During the use of the AI model, we keep monitoring the performance and operation of the AI model. Many AI frameworks support visualization tools to analyze the AI model. Thus, the AI model can be updated while observing the performance change.

AI techniques will be useful to improve the performance of 6G systems, develop a new application, and optimize 6G systems. The key applications of AI techniques are classification, clustering, prediction, dimension reduction, and decision making. Many functions of 6G systems are in line with the key applications of AI techniques. For example, channel estimation of physical layer is about interpolation and prediction. The channel problem can be formulated as a prediction problem. The prediction can be solved by AI techniques such as gradient decent algorithm, deep learning, and so on. In addition, network traffic prediction is directly related to AI techniques such as DNN. The mobile networks should be monitored to detect network faults, distributed denial-of-service (DDoS) attack, elephant traffic, and so on. The prediction function of the AI techniques will be able to improve the network performance. However, there are many research challenges to implement AI techniques in 6G systems. We can summarize the key questions as follows: (1) How do we acquire the training dataset with suitable features? (2) How do we select or develop an AI algorithm to match well with the functions of cellular communications and networks? (3) What network parameters (such as throughput, latency, energy efficiency, connection density, and so on) do we improve for specific services? (4) How do we develop the AI model when wireless channels are varying (Namely, coherent time is shorter than the channel bandwidth)? (5) How do we find a common theoretical background to implement both AI theory and wireless communication theory? (6) Can we develop an optimization theory to design 6G systems embedded 6G systems?

Summary 3.3 AI Workflow and Implementation in 6G

1. The overall workflow of AI implementation is to collect data, create a training model to learn a pattern in training data, train a model, validate a trained model, find the pattern in other data, monitor the performance of the model and manage the model.
2. The key applications of AI techniques are classification, clustering, prediction, dimension reduction and decision making. Many functions of 6G systems are in-line with the key applications of AI techniques.

References

1. Haesik Kim, Artificial intelligence for 6G, Springer, (2022), ISBN-10: 3030950409.
2. John McCarthy, Marvin L. Minsky, Nathaniel Rochester, Claude E. Shannon, A proposal for the dartmouth summer research project on artificial intelligence, August 31, 1955, AI Mag., **27**(4), p. 12, (2006)
3. John R. Searle, Minds, brains, and programs. Behav. Brain Sci., **3**(3), pp.417–424, (1980)
4. James Lighthill, Artificial intelligence: a general survey. In Artif. Intell.: Pap. Symp., Sci. Res. Counc., (1973)
5. D.H. Wolpert, W.G. Macready, No free lunch theorems for optimization. IEEE Trans. Evol. Comput., **1**(67), (1997)
6. W.S. McCulloch, W. Pitts, A logical calculus of the ideas immanent in nervous activity. Bull. Math. Biophys. **5**, 115–133 (1943). https://doi.org/10.1007/BF02478259
7. F. Rosenblatt, The perceptron: A probabilistic model for information storage and organization in the brain. Psychol. Rev. **65**(6), 386–408 (1958). https://doi.org/10.1037/h0042519
8. D. Rumelhart, G. Hinton, R. Williams, Learning representations by back-propagating errors. Nature **323**, 533–536 (1986). https://doi.org/10.1038/323533a0

Internet of Things—An Introduction

4

4.1   Introduction to IoT

Back in 1990, John Romkey, a visionary American computer scientist, marked a technological milestone by introducing a smart device that was considered the first of a kind IoT appliance, the Internet-enabled toaster [1]. This seemingly mundane household appliance paved the way for a new era of the internet based technological transformation. This innovation marked a significant milestone in the development of the Internet of Things paradigm that brought together the physical and digital worlds by connecting everyday devices to the internet. The advanced embedded systems and programming of this device allowed it to interact with other devices, revolutionizing the way we think about ordinary objects and the role they play in the digital landscape. This event demonstrated that it was not only technically impressive but also symbolic of the tremendous potential that exists in the intersection of technology, creativity, and convenience.

The growth of IoT is expected to escalate dramatically in the coming years, with the number of connections on the network predicted to increase from 15.1 billion in 2021 to over 23 billion in 2025, according to a report from the Global System for Mobile Communications Association (GSMA) [3]. Before the advent of IoT, the internet was limited to desktops, laptops, and smartphones, and the network connecting these devices was defined as the Internet of Content. The era of the Internet of Content began with the exploration of emails and messages before shifting to the Internet of Services, which included applications like e-commerce. From there, it became the Internet of People, allowing people to communicate through social media platforms such as Facebook and Skype [4, 6, 7].

The Internet of Things takes this connectivity to the next level, allowing all devices ranging from televisions to watches, from refrigerators to cars, and even washing machines to be connected to the internet [8–10]. These interconnected devices have the

© The Author(s), under exclusive license to Springer Nature Switzerland AG 2025
V. Sai et al., *6G Enabled Healthcare Systems*, Synthesis Lectures on Engineering,
Science, and Technology, https://doi.org/10.1007/978-3-031-73849-4_4

capability to communicate with each other and perform specific functions that require human inputs. With the integration of AI and IoT, the current phase of IoT research has seen extensive progress in enabling connected devices to perform actions without human involvement. The possibilities are endless with IoT and AI, with technology now working closely with humans to make tasks easier and better [11–13].

The IoT has transformed technology as we know it, creating a network of devices that communicate with each other over wireless networks, completely independent of human intervention. The examples of IoT are endless, with everything from air conditioners to doorbells now being able to communicate with each other, paving the way for a smarter and more efficient future. In other words, IoT has evolved to make connections between people and devices possible at any time, in any place, and with anyone, anywhere, utilizing any network and any service that is available, as depicted in Fig. 4.1 [5].

The Internet of Things ecosystem is a global network that enables the dynamic interconnection among the physical, digital, and virtual worlds. This interconnected network connects individuals, objects, data, services, processes, and more, elevating the quality of life and work in unparalleled ways. Technology has endless applications in providing healthcare, optimizing logistics, improving transportation and supporting a sustainable and resourceful future, among other areas [6]. IoT technology is flexible and scalable, capable of communicating, collecting and analyzing data in real time, enabling meaningful insights to support decision making processes [5, 11–13]. Innovations in IoT

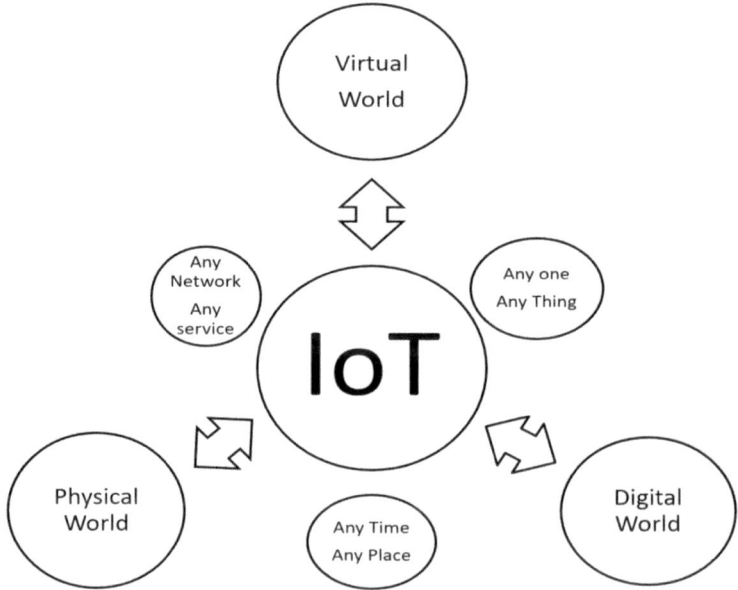

Fig. 4.1 IoT ecosystem concept

technology are rapidly transforming the way we live and work, ushering in a new era of interconnectedness poised to revolutionize the world.

Summary 4.1

1. The primary mission of IoT is to create an ecosystem of intelligent systems that can communicate with each other and actively participate in the decision-making process.
2. Ultimately, the result of IoT is to make life easier and more meaningful for people by automating complex systems and processes, thus enabling users to concentrate more on things they care most about.

4.2   How Does IoT Work?

The operational landscape of IoT networks involves a vast array of devices and sensors that work together in real-time to detect and gather relevant data. These data are then channelled through communication media, such as gateways, and forwarded to storage and processing components, which can be cloud-based services or other platforms that store and process data [11, 12]. This initial data processing stage is where IoT data really start to become valuable.

Once the data have been collected, it undergoes a comprehensive analysis and processing using advanced AI and ML techniques to extract valuable insights. The AI and ML mechanism enables the system to make intelligent deductions based on the data collected, thus enabling them to arrive at conclusions. These intelligent deductions could range from providing real-time monitoring data to detecting patterns and implicit messages from data collected from the IoT devices [11–13].

During this processing phase, the system analyzes the data to discover insights, determine new trends, and make predictions. All of these processes help the system to make intelligent decisions that are then sent back to the user interface for further action and implementation. This two-way communication between the system and the user is essential because it allows the user to interact with the data and gain value and insights, while the system learns more about the user's preferences and priorities. This interaction between users and the system leads to an increase in understanding of IoT devices, which can be utilized to simplify complex systems and enhance the user's quality of life.

IoT Architecture has the following basic elements as shown in Fig. 4.2 [2, 14, 15].

(a) Sensing and or Device element: The 'thing' component of the IoT architecture functions as a pivotal element. This architecture facilitates the conversion of physical inputs that originate from the environment, such as sound, light, and other such inputs, into digital signals that can be processed by the computing system. The underlying

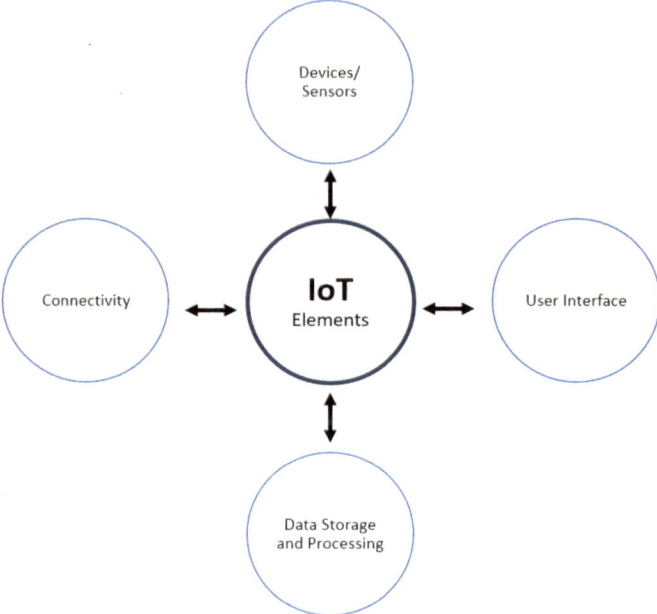

Fig. 4.2 Basic elements of the IoT system [2]

mechanism includes the collection of data inputs by specialized sensors and devices placed in the surrounding areas. These sensors are designed to capture specific data inputs related to distinct functions such as temperature, light, motion detection, atmospheric pressure, smoke, and other relevant parameters. These components play a crucial role in acquiring and delivering accurate data inputs, which eventually lead to processing and analysis through the computing system for the generation of valuable insights that could be used to make informed decisions in various settings.

(b) Connectivity element: After the data are collected using various sensors and devices, they are transmitted to the cloud using a variety of communication technologies such as Bluetooth, WiFi, RFID, NFC, or any other relevant protocols. These communication technologies are essential for enabling a seamless flow of data from sensors to the cloud. The availability of these technologies broadens the scope of IoT-enabled applications, facilitating the creation of numerous innovative solutions in various domains. Figure 4.3 lists some of the primary communication technologies available for IoT, all of which are essential for establishing uninterrupted connectivity, and ensuring a reliable flow of information from one point to another [4, 12, 15].

(c) Data Storage and Processing element: After the data collected from various sensors reaches the cloud, advanced software analytics are employed to extract meaningful insights and patterns related to the collected data. These insights are then utilized to

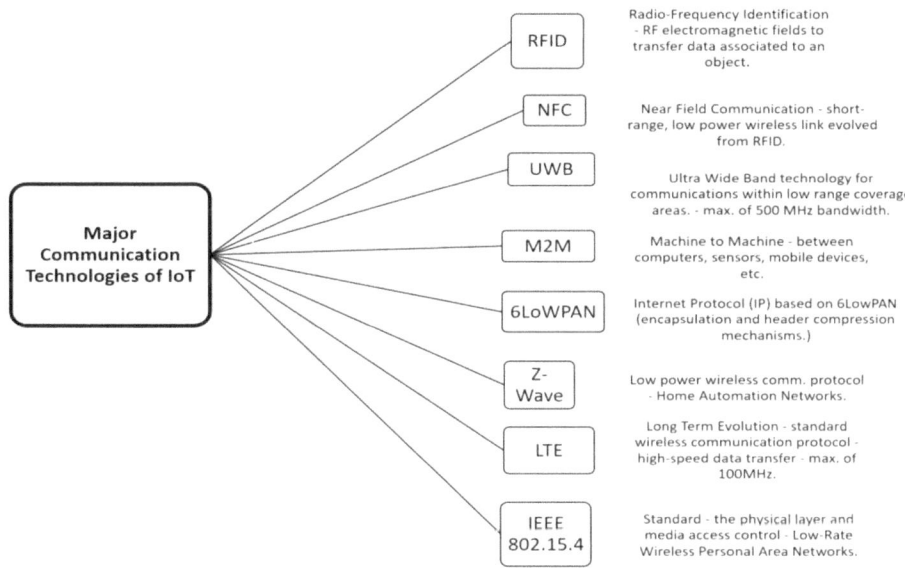

Fig. 4.3 Communication technologies of IoT

make intelligent decisions that can improve the system's performance, efficiency, and accuracy. Once the decision-making process is complete, the system passes on the decision to the respective device for further action and implementation. The reflexive decision-making process has immense potential in various domains and can enable the deployment of highly sophisticated and intelligent systems capable of making autonomous and informed decisions. Therefore, through the process of data analytics and advanced algorithms, IoT systems can deliver high-quality insights that can drive continuous improvements and enable better performance.

(d) User interface element: Interaction is a critical aspect of IoT devices, and they are designed to cater to the demands and preferences of the user. This interface acts as the front-end layer that provides users with the necessary tools and controls required to interact with the back-end elements of the IoT architecture. This layer is typically the most accessible part of the IoT system, making it easy for users to engage with the system and obtain insights based on their specific requirements. By providing intuitive interfaces, IoT systems can be more user-friendly, allowing users to customize settings based on their preferences and allowing them to access the necessary information at their convenience. Ultimately, the user interface is an essential element of the IoT ecosystem, enhancing the user experience and ensuring better outcomes.

Summary 4.2

1. Any device or appliance, from your refrigerator to your home security system, can be connected to the internet, thus functioning as an IoT device.
2. IoT is an ecosystem of internet-connected devices that collect and share data, using sensors and advanced technologies for real-time processing and analytics using AI and ML techniques to extract insights and make intelligent decisions.
3. The IoT architecture has four basic elements: sensing and device, connectivity, data storage and processing, and user interface.
4. These elements work together to convert physical inputs into digital signals, transmit data to the cloud, extract insights through analytics, and provide a user-friendly interface for interacting with and customizing the system.

4.3   Why Do We Need IoT?

The Internet of Things (IoT) has become a network of every possible device that can connect to the internet, collect, and share data. From smartphones to smart home appliances, from health monitoring devices to autonomous vehicles, everything that is connected to the internet forms a part of the IoT ecosystem.

IoT systems are designed to collect data from various sources using sensors, actuators, and other cutting-edge technologies. These data are then shared with decision-making systems, which use automation, analytics, machine learning, and AI, among other techniques, to process this information in real time. The ability of IoT systems to collect and analyze data in real time is one of their biggest advantages, enabling them to deliver accurate, reliable, relevant, and timely information for better decision-making.

IoT systems make use of networking, sensing, big data, AI, and other technologies to drive greater transparency, control, and performance in the delivery of complete systems for a product or a service. These systems are capable of monitoring, analyzing, and controlling various aspects of the product or service, making it possible to improve performance, optimize efficiency, and reduce the possibility of errors.

Thus, the Internet of Things has revolutionized the way devices connect and interact with each other, enabling the creation of a seamless and integrated ecosystem of smart devices that can collect, share, and process data to deliver real-time insights and actionable outcomes. The scope of IoT is expanding rapidly, with new devices and applications being introduced every day, promising to make our lives more connected, informed, and efficient.

In today's world, where the Internet of Things (IoT) has become incredibly pervasive, the flow of data has become a crucial factor in the delivery of accurate, relevant, dependable, and timely information. To understand the various facets of IoT data flow, it is essential to look at Fig. 4.4, which represents a typical IoT data flow [10, 12, 13].

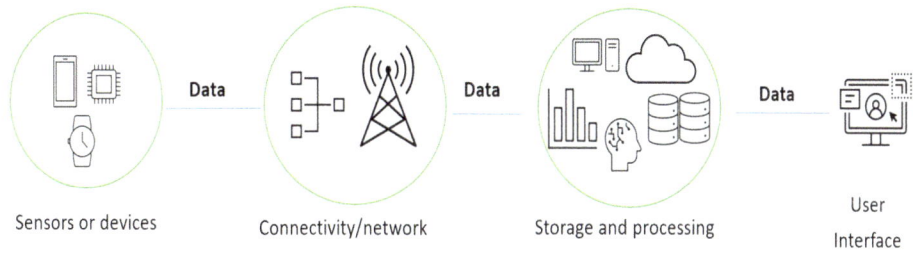

Sensors or devices   Connectivity/network   Storage and processing   User Interface

Fig. 4.4 Data flow in an IoT system

As depicted in the figure, the processing and cloud components of the flow are where the automation and advanced analytics on the data take place. These storage and computational entities form the backbone of the IoT ecosystem, enabling the integration of data from multiple sources, thus facilitating better decision making.

By leveraging advanced analytics and automation systems, industry players across all sectors can gain deeper insights into the performance of their products and services. By processing data in real-time, they can identify trends, predict future outcomes, and make informed decisions that can help them stay ahead of the competition. The deployment of cloud computing and storage technology in these storage and computational systems has also made it possible to process and analyze vast amounts of data generated by IoT devices [10, 12, 13].

The integration of data through cloud computing and storage systems enables the deployment of advanced analytics and automation, which helps to achieve greater transparency, control, and performance when applied to any industry. The volumes of data generated by IoT devices are enormous, and it is impossible to process them efficiently without the use of advanced technologies. However, with the support of cloud computing and storage technology, it is now possible to process and analyze data in real-time, unlocking the full potential of IoT.

In essence, the IoT data flow is a critical factor in the delivery of accurate, relevant, reliable, and timely information. The processing and cloud components of the flow are where the automation and advanced analytics of the data take place. The integration of data through cloud computing and storage systems can help organizations achieve greater transparency, control, and performance when applied to any industry. Through the application of advanced analytics and automation, organizations can gain valuable insight and make informed decisions that can help them stay ahead of competition and drive growth and positive results.

The Internet of Things (IoT) technology has revolutionized the way technology works, and it has continued to evolve ever since its inception. The latest advancements in IoT include the incorporation of cutting-edge technologies such as the fifth and sixth generation (5G/6G) connectivity solutions and Artificial Intelligence (AI) and Machine Learning

(ML) applications. The infusion of these disruptive technologies has resulted in better connectivity and a significant decrease in human intervention in IoT-enabled systems, thereby providing authorized access to the systems from anywhere and at any time.

This fusion of advanced technologies has also created new opportunities for the development of context-aware data. Such data can help develop autonomic IoT applications in areas such as smart transportation and smart cities, among others [16–18]. These applications can have a significant impact on the way people live, work and interact with technology.

Smart transportation, for example, can use real-time data collected from IoT-enabled devices to optimize traffic flow, provide accurate traffic predictions, and ultimately reduce traffic congestion. Context-aware data, combined with AI/ML applications, can help vehicles make decisions based on real-time traffic conditions and optimize the efficiency of transportation networks.

Smart cities, on the other hand, can leverage IoT technology to improve energy efficiency, reduce waste and environmental impact, enhance safety and security, and provide a better quality of life for citizens. The incorporation of AI/ML solutions into IoT-enabled systems can help achieve these goals by analyzing large volumes of data generated by IoT devices, thus enabling the development of intelligent systems that can make decisions and take action autonomously.

In conclusion, the evolution of IoT technology has resulted in the integration of advanced technologies, including 5G/6G, AI, and ML, leading to better connectivity, reduced human intervention, and increased access to IoT-enabled systems. The availability of context-aware data resulting from these technologies has enabled the development of autonomic IoT applications in areas such as smart transportation and smart cities, which have the potential to transform the way we live, work, and interact with technology [17, 18].

Summary 4.3

1. Cloud computing and storage technology support IoT systems by integrating data from multiple sources, allowing for automation and advanced analytics that provide actionable insights and informed decision-making.
2. The latest advancements in IoT include the infusion of disruptive technologies like 5G/6G, AI, and ML, which have created opportunities for the development of context-aware data, leading to the development of autonomic IoT applications in areas such as smart transportation and smart cities that have the potential to transform our way of living and working.

4.4   Applications of IoT

IoT has evolved to become much more than just a system that interconnects 'things' that are connected to the internet. Instead, IoT has taken every aspect of life and made it 'smart' by leveraging the power of data collection, advanced algorithms, networks, and AI/ML. Figure 4.5 represents various use cases of IoT. These use cases encompass consumer applications, organizational applications, industrial applications, and infrastructure applications [19].

Consumer applications of IoT are perhaps one of the most visible examples of this technology in the current day and age. Smart homes, connected vehicles, and wearables are some popular applications of the IoT in consumer technology. In smart homes, IoT sensors are used to automate daily tasks such as lighting, temperature control, and security. Connected vehicles, on the other hand, are equipped with connected technologies that allow for real-time monitoring of driving data, fluctuations in traffic, speed, etc. Wearables, such as fitness trackers, smartwatches, etc., are other popular applications of IoT.

Organizational applications of IoT are aimed at improving the overall efficiency of the organization. Smart healthcare and smart transportation are two such examples of IoT

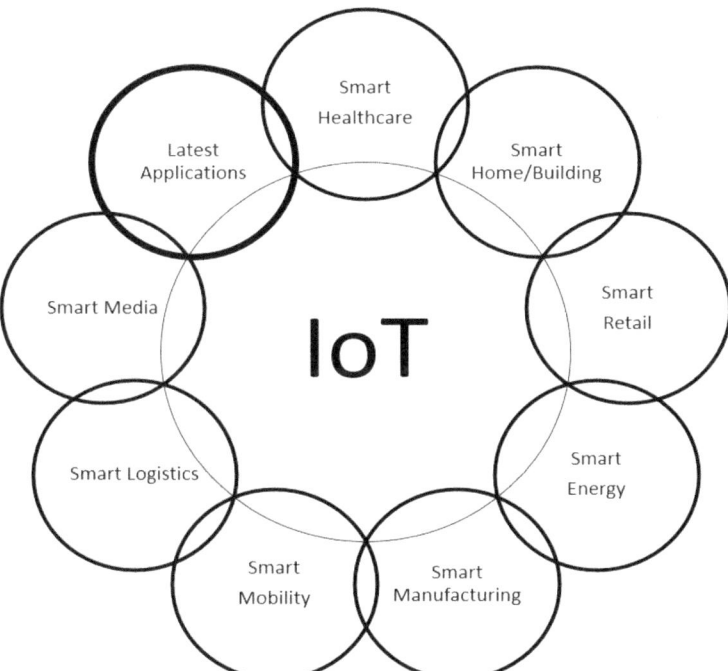

Fig. 4.5 Applications of IoT systems [2]

applications [19]. IoT in healthcare is specifically designed to provide better and more personalized medical care to patients by leveraging the power of the internet. Transportation systems, on the other hand, are designed to provide real-time transportation monitoring, thus reducing congestion, improving safety, and providing better overall mobility.

In industrial settings, IoT has brought about tremendous changes by enabling smart manufacturing, agriculture, and logistics management [19, 22]. IoT-based logistics and supply chain management systems have helped reduce the duration of the delivery process, minimize errors, and optimize inventory management. Smart agriculture systems incorporating IoT tech have enabled better farm management practices, significantly improving crop yields [20].

Finally, the Internet of Things has also transformed cities into smart cities, where various municipal services are connected and simplified to improve quality of life and reduce costs. Smart city applications include traffic management and transportation, waste management, energy and water management, public safety and security, and much more [19, 21].

Ultimately, IoT has emerged as a game changer in various industries. With its potential to reduce human involvement, boost efficiency, and offer economic benefits, the application of IoT continues to grow, with an ever-expanding list of use cases that have transformed our everyday lives. The bold circle in Fig. 4.5 represents all the other current and ever-expanding IoT applications and possibilities available to us.

The advent of IoT has revolutionized the way people live their lives, and one of the most popular consumer applications of IoT today mentioned earlier is smart homes. At the individual level, smart homes are built on a platform that allows the control of various smart devices using sensors. This technology has been designed to create seamless experiences for consumers, while improving their efficiency and safety as well.

One of the features that has become quite popular within smart homes is automatic lighting, where lights can automatically detect the presence of occupants within the room and activate or deactivate lighting accordingly. For added convenience, voice activation features are also popular in smart homes, allowing users to control different aspects of their homes with their voice as well. Smart homes have also been integrated with advanced temperature management features that function based on the surrounding climate to automatically control the temperature of the house.

Security measures are another popular feature of smart homes. These advanced security systems are equipped with various components such as cameras, sensors, and alarms that support the purpose of keeping homes safe. They are built with features that allow for the locking of doors, windows, and gates once the occupant leaves the house. This integration of security and automation within smart homes designed for individual use has changed the way people look at the concept of 'home-sweet-home.'

Smart homes are integral components of an even bigger system—the smart city. They are envisioned as an independent technological entity aimed to provide a comprehensive ecosystem of modern living, which is expected to blend seamlessly with other parts of the city infrastructure. The idea is to create a broader network across the city where all these independent technologies, such as traffic management and transportation systems, can be synchronized to work together to increase the overall efficiency and quality of city services.

In the end, smart homes are here to stay. They are designed to provide comfort, energy savings, security, safety, and convenience while seamlessly integrating within the broader concept of a smart city, aiming to improve city living at the individual and community level. These technological advancements will continue to evolve, enhancing lifestyles, and the general quality of life for the masses.

The healthcare sector is constantly advancing with the integration of IoT and in an ever-evolving industry, its importance only continues to grow further. The use of IoT in healthcare also has multiple organizational applications. Perhaps, one of its most significant contributions has been in the development of remote healthcare systems and emergency notification systems. This has extended its capabilities beyond just monitoring and tracking, as it enables remote interaction between doctors and patients, providing access to medical attention anytime and anywhere, regardless of distance.

Patient data analysis aims to make healthcare delivery more personalized and effective by aggregating patient data to monitor their medical histories and relevant information. Data obtained from such analyses is essential to improve healthcare outcomes and has the potential to revolutionize the healthcare industry with better patient outcomes.

Additionally, the applications of IoT do not stop there, as it also has several other applications. Smart parking is another IoT application that has made parking's a more seamless process. The technology used in its implementation includes the use of sensors that track parking patterns to make recommendations for nearby vacant parking spots for drivers, reducing time spent searching for available parking, and resulting in fewer traffic and parking disruptions.

In a nutshell, IoT has immense applications in the healthcare sector, with its capabilities extending well beyond traditional monitoring and tracking services. It redefines the entire healthcare delivery system and creates opportunities for improved patient outcomes and medical services. Similarly, with the integration of IoT in smart parking, it has become possible for drivers to find vacant parking spots with ease, reducing the time spent searching and the traffic congestion that comes with it. This proves that IoT continues to push boundaries, creating innovative solutions for everyday challenges.

Smart Manufacturing is a popular industrial Internet of Things application that has significantly changed the manufacturing landscape globally. This revolutionary technology enables machinery with Internet-enabled monitoring to automatically supervise the entire manufacturing process. The entire production process, including the ordering of raw materials as they decrease, is closely monitored, increasing productivity, efficiency, and streamlines the entire workflow. This is an excellent way of ensuring that the manufacturing process remains seamless while halting the production process in case of any anomalies or issues occurring in the process.

Complementing smart manufacturing is the smart agriculture application, another key IoT area that has immense implications for food production. Modern technology utilizes a wide range of IoT-based devices that can collect soil data, such as its composition, temperature, and other critical parameters such as pest infestation. This information helps farmers to obtain data-based insights into the crop requirements, the availability of nutrients in real-time, etc. This enables farmers and other agriculturists to make more informed decisions when it comes to crop scheduling, fertilizing, and pest and weed control to ultimately deliver better yields. This plays an essential role in providing an adequate food supply to the growing population that is consistently on the rise. Industries are beginning to pay more attention to these areas and develop comprehensive IoT solutions for our convenience.

The concept of IoT has emerged as one of the most significant drivers of modern technology, enabling the integration of the physical world with digital computing systems. One of the most important applications of IoT includes monitoring and controlling infrastructures in smart cities. Bridges, railway tracks, and other such infrastructures require regular maintenance and monitoring to maintain optimal functionality and ensure the safety of citizens.

In this context, IoT has emerged as a game-changing solution for smart cities, providing a comprehensive range of tools and technologies to monitor and control the functioning of infrastructures. For instance, IoT-enabled sensors can be deployed on bridges and other such structures, which can monitor vibrations, temperature, and other similar parameters in real-time. This data can be analysed using machine learning algorithms to predict the likelihood of breakdowns and initiate corrective action before the fault occurs.

Similarly, railway tracks can also be monitored using IoT-enabled solutions to predict and preempt train accidents. IoT sensors can be deployed at various points along the tracks to detect changes in temperature, sound, vibrations, and other such parameters. These data can be analyzed in real time using advanced analytics tools to predict potential faults and address them before they cause any damage.

In general, the power of the IoT has been demonstrated through its ability to improve the efficiency and productivity of various ecosystems. The use of IoT in smart cities has made it possible to monitor and control critical infrastructure, thereby ensuring the safety and security of citizens. Using IoT solutions, the physical world can be integrated with digital computing systems to achieve significant gains in productivity and efficiency, leading to new opportunities for growth and development.

Summary 4.4

1. IoT has various use cases encompassing consumer, organizational, industrial, and infrastructure applications, with examples such as smart homes, smart healthcare, smart agriculture, and smart cities.
2. IoT has the potential to reduce human involvement, enhance efficiency, and offer economic benefits, transforming the way people live their lives, and influencing industries from manufacturing to healthcare and infrastructure.

4.5 IoT Challenges

As IoT technology gains momentum and continues to evolve, it brings with it a host of new and innovative applications that can transform the way we live and work. Examples of these groundbreaking applications include smart homes, wearable technology, connected cars, smart cities, and many others. However, as with any rapidly evolving technology, the IoT poses its own set of challenges that must be addressed to ensure its continued growth and success [19, 23, 24].

One of the most significant challenges facing the ever-expanding world of IoT is security. With millions of connected devices transmitting data over the internet, the risk of data breaches has never been higher. IoT devices often carry sensitive information such as personal data, financial information, and confidential business data, making them vulnerable targets for cybercriminals. To address this issue, security protocols must be developed and implemented at all levels of IoT infrastructure.

Another challenge associated with IoT is the need for interoperability and standardization. As new devices and applications are created, they must be able to communicate with other devices and systems effectively. This requires interoperability standards to be put in place to ensure that devices can communicate across different platforms and operating systems.

The scalability of IoT applications is yet another challenge that must be addressed. As the number of connected devices increases, it becomes difficult to manage and control these devices centrally. This requires advanced analytics tools and automation systems to be put in place to ensure that IoT systems can scale as needed.

Finally, the issue of data privacy is a critical challenge that IoT system designers must address. With the growing number of connected devices transmitting data over the internet, the privacy of this sensitive data must be protected. This requires that encryption protocols and other security measures be implemented at all levels of the IoT infrastructure.

Although IoT technology has brought with it a world of new applications and possibilities, it also poses its own set of challenges. These include security, interoperability and standardization, scalability, and data privacy. To ensure IoT's continued growth and success, these challenges must be addressed proactively and systematically.

The proliferation of IoT technology has led to the collection and sharing of vast amounts of data, providing new and innovative ways to monitor and control various systems. However, this vast amount of data also raises concerns over privacy and security, particularly in critical applications such as healthcare, where sensitive data is often involved. The need to ensure the privacy and security of user data has become a critical area of concern for developers working in the IoT ecosystem.

In IoT systems, data is collected from a variety of sources and often transmitted to various intermediate layers for processing and analysis. The complex nature of this data flow makes it challenging to maintain data confidentiality and security. For example, in healthcare applications, patient data, including medical history, treatment plans, and other sensitive information, are collected and shared with various stakeholders, such as doctors, nurses, and healthcare providers. As this data is transmitted through various layers of the IoT ecosystem, it is essential to ensure proper data encryption and authentication protocols to protect sensitive patient information.

Another significant security concern in IoT systems is the risk of cyberattacks. Cybercriminals can gain unauthorized access to sensitive data and exploit vulnerabilities in the system to cause damage or disrupt operations. To prevent this, security protocols must be implemented at all levels of the IoT infrastructure, including device authentication, data encryption, and user access control.

Furthermore, as IoT-enabled devices become more ubiquitous, the need for increased regulatory oversight and compliance has become increasingly relevant. Regulations are needed to ensure that IoT systems adhere to established security standards and protocols, protecting user privacy and preventing unauthorized access to sensitive data.

In summary, privacy and security concerns in IoT applications are of critical importance and need to be addressed proactively. Sensitive data must be protected from unauthorized access and breaches, and security protocols must be implemented at all levels of IoT infrastructure. With proper regulations and security measures in place, the promise of IoT technology can be realized without compromising user privacy or data security.

As the world becomes increasingly dependent on IoT devices, there has been a growing demand for intelligent devices that can perform complex tasks on the edge of networks. However, this demand has raised concerns over energy efficiency. Many of these devices have limited power or must be continuously powered to perform processing tasks and transmit data to the cloud. This requirement has led to a significant challenge for developers who must balance the processing power of IoT devices with their power consumption.

To address this challenge, researchers around the world have been working to develop new technologies that can improve the energy efficiency of IoT devices. One approach is to use low-power computing platforms that consume less energy than traditional processors. Many companies have made significant strides in this area by developing specialized processors, such as Arm's Cortex-M series and Intel's Quark processors, that are specifically designed for low-power IoT applications.

Another strategy to improve energy efficiency is to use edge computing. With edge computing, data are processed and analyzed locally, rather than being transmitted to the cloud, reducing the amount of data that needs to be transmitted and ultimately reducing power consumption. Some new IoT devices can perform edge computing by processing data from sensors locally, which means that data do not need to be transmitted to the cloud for every computation.

In addition to low-power computing platforms and edge computing, other techniques, such as power management and optimization algorithms, can also be used to improve energy efficiency. As devices become more advanced, they require more power, which can be challenging in contexts where power is limited or unreliable. To address this issue, developers are designing devices with low-power modes, which can be used to reduce power consumption when the device is not in use. Furthermore, other power-saving techniques, such as sleep modes, can be used to reduce power consumption during periods of low activity. These techniques can aid in reducing the power consumption of IoT systems by maximizing energy efficiency and minimizing the amount of unnecessary processing.

While the need for intelligent IoT devices on the edge of networks continues to grow, the energy efficiency of these devices must also be considered. Developers must balance processing power with power consumption, which can be achieved by using low-power computing platforms, edge computing, or power management and optimization algorithms. As these technologies continue to evolve, we can expect even more energy-efficient IoT devices that can perform complex tasks while consuming minimal power.

With the ever increasing number of IoT devices available today, it is becoming increasingly important that these devices work together seamlessly to accomplish relevant tasks. Cross-compatibility, inter-operability, and standardization have become the cornerstones of this new era of interconnected devices. By building devices that can communicate with each other, developers can create more complex and useful systems that can provide a whole new level of functionality and data analysis.

One area where these challenges have become particularly important is in the area of data management. With the rise in the number of devices, there is also a proportional increase in the volume of data produced by these devices. Managing this data requires advanced algorithms and infrastructure to ensure that data is collected efficiently, accurately, and securely. In addition, devices must also be designed to manage and analyze data in real-time, which often requires significant computational power.

Another important challenge when it comes to IoT devices is network connectivity. IoT devices must maintain uninterrupted network connectivity to operate effectively, which is not always possible due to issues such as network congestion, signal interference, and signal loss. To address these issues, developers are designing devices with built-in redundancy and failover mechanisms, ensuring that data is always transmitted even if the network connection is lost.

In conclusion, the growing number of IoT devices available today presents new challenges that must be addressed to ensure success in this space. Cross-compatibility, inter-operability, and standardization are crucial, along with data management, network connectivity, and power management. As devices continue to evolve and become more advanced, these challenges will continue to drive innovation in the IoT ecosystem, leading to new and exciting applications of this technology in various industries.

Summary 4.5

1. IoT technology offers innovative applications such as smart homes, wearable technology, connected cars, smart cities, and more, but also presents significant challenges, including security, interoperability, scalability, and data privacy.
2. IoT system designers must prioritize privacy and security due to the vast amount of sensitive data being collected and transmitted, which requires proper encryption, authentication, and user access control protocols.
3. Developers are working on improving energy efficiency by using low-power computing platforms, edge computing, power management and optimization algorithms, and other techniques, balancing processing power with power consumption.
4. Additionally, cross-compatibility, inter-operability, and standardization are crucial aspects of IoT device operation that require advanced algorithms and infrastructure, while network connectivity and power management present additional challenges that need to be addressed for IoT's continued success.

References

1. J. Romkey, Toast of the IoT: the 1990 interop internet toaster. IEEE Cons. Electron. Mag. **6**(1), 116–119 (2017). https://doi.org/10.1109/MCE.2016.2614740
2. R. Minerva, A. Biru, D. Rotondi, Towards a definition of the internet of things (iot). IEEE Internet Initiat. **1**(1), 1–86 (2015)
3. https://www.gsma.com/mobileeconomy/wp-content/uploads/2022/02/280222-The-Mobile-Eco nomy-2022.pdf
4. A. Khanna, S. Kaur, Internet of things (IoT), applications and challenges: a comprehensive review. Wirel. Pers. Commun. **114**, 1687–1762 (2020). https://doi.org/10.1007/s11277-020-074 46-4
5. A. Ghasempour, Internet of things in smart grid: architecture, applications, services, key technologies, and challenges. Inventions **4**, 22 (2019). https://doi.org/10.3390/inventions4010022

6. H. Liu et al., A review of the smart world. Future Gener. Comput. Syst. 96, 678–691 (2019). https://linkinghub.elsevier.com/retrieve/pii/S0167739X17319532

7. S. Dhelim, H. Ning, N. Aung, ComPath: user interest mining in heterogeneous signed social networks for internet of people. IEEE Internet Things J 8(8), 7024–7035 (2021), https://doi.org/10.1109/JIOT.2020.3037109

8. M. Domb, Smart home systems based on internet of things, in *IoT and Smart Home Automation* (IntechOpen, 2019)

9. O. Taiwo, A.E. Ezugwu, Internet of things-based intelligent smart home control system. Secur. Commun. Netw. Article ID 9928254, 17 (2021)

10. L. Babangida, T. Perumal, N. Mustapha, R. Yaakob, Internet of things (IoT) based activity recognition strategies in smart homes: a review. IEEE Sens. J. 22(9), 8327–8336 (2022)

11. Y.A. Qadri, A. Nauman, Y.B. Zikria, A.V. Vasilakos, S.W. Kim, The future of healthcare internet of things: a survey of emerging technologies. IEEE Commun. Surv. Tuts. 22(2), 1121–1167, 2nd Quart., (2020)

12. Z. Chang, S. Liu, X. Xiong, Z. Cai, G. Tu, A survey of recent advances in edge-computing-powered artificial intelligence of things. IEEE Internet Things J. 8(18), 13849–13875 (2021)

13. S. Baker, W. Xiang, Artificial intelligence of things for smarter healthcare: a survey of advancements, challenges, and opportunities. IEEE Commun. Surv. Tutor. 25(2), 1261–1293, Secondquarter (2023)

14. L. Chettri, R. Bera, A comprehensive survey on internet of things (IoT) toward 5G wireless systems. IEEE Internet Things J. 7(1), 16–32 (2020)

15. G.A. Akpakwu, B.J. Silva, G.P. Hancke, A.M. Abu-Mahfouz, A survey on 5G networks for the internet of things: communication technologies and challenges. IEEE Access 6, 3619–3647 (2017)

16. A. Al-Fuqaha, M. Guizani, M. Mohammadi, M. Aledhari, M. Ayyash, Internet of things: a survey on enabling technologies, protocols, and applications. IEEE Commun. Surv. Tuts. 17(4), 2347–2376, 4th Quart. (2015)

17. C.K. Toh, J.A. Sanguesa, J.C. Cano, F.J. Martinez, Advances in smart roads for future smart cities. *Proceedings of the Royal Society A: Mathematical, Physical Engineering Science* 476 (2020)

18. M. Tondro, M. Jahanbakht, Utilizing IoT technological innovation by startup businesses for sustainable smart transportation in developing countries, in *2023 IEEE Conference on Technologies for Sustainability (SusTech), Portland, OR, USA* (2023), pp. 293–298

19. D.C. Nguyen, M. Ding, P.N. Pathirana, A. Seneviratne, J. Li, D. Niyato, O. Dobre, H.V. Poor, 6G internet of things: a comprehensive survey. IEEE Internet Things J. 9(1), 359–383 (2022)

20. P. Tokekar, J. Vander Hook, D. Mulla, V. Isler, Sensor planning for a symbiotic UAV and UGV system for precision agriculture, in *Proceedings of the 2013 IEEE/RSJ International Conference on Intelligent Robots and Systems, Tokyo* (2013), pp. 5321–5326

21. S. Kisseleff, W.A. Martins, H. Al-Hraishawi, S. Chatzinotas, B. Ottersten, Reconfigurable intelligent surfaces for smart cities: research challenges and opportunities. IEEE Open J. Commun. Soc. 1, 1781–1797 (2020)

22. Y. Xiao, G. Shi, Y. Li, W. Saad, H.V. Poor, Toward self-learning edge intelligence in 6G. IEEE Commun. Mag. 58(12), 34–40 (2020)

23. S. Nayak, R. Patgiri, 6G communication technology: a vision on intelligent healthcare. IEEE Internet Things J. 7(10), 9227–9234 (2020)

24. S.N. Nova, M.S. Rahman, C. Chakraborty, 'Patient's health surveillance model using IoT and 6G technology,' in *Green Technological Innovation for Sustainable Smart Societies*. ed. by C. Chakraborty (Springer, Cham, Switzerland, 2021), pp.191–209

IoT for Healthcare Systems: Use Case 6G-IoT Application

The healthcare industry has undergone significant changes and improvements over time due to continuous innovation and the ever-increasing advancement of information and communication technologies (ICT). Through the years, ICT has revolutionized the healthcare sector, transforming the way healthcare services are provided, delivered, and managed. Although the progress of this sector dates to the late 1990s, it has only recently begun to unleash its full potential with the advent of Industry 4.0.

Industry 4.0 includes revolutionary technologies such as AI, big data, and robotics, among others that has given rise to a new era of healthcare [1]. Healthcare in digital form is the new age of digital health. Figure 5.1 illustrates how health information technology has progressed over time, leading us to the current state of digital health. This new concept is all-encompassing and includes various significant sectors such as telemedicine, telehealth, mobile health, health information technology, wearable health monitors, and much more [2].

As discussed previously, IoT technology has a fundamental and critical role to play in the healthcare industry, with the potential to save not only time and money but also lives. This chapter will delve into the ways in which IoT is revolutionizing the healthcare industry and the ways it is improving patient lives globally. Due to the interconnected nature of IoT, patients are becoming more connected and in turn generating more data. This leads to clinicians being able to identify and address their specific needs more efficiently and accurately. The potential for personalized preventive care and other innovations in healthcare is limitless, especially due to the latest advanced data science techniques.

Therefore, IoT technology is unveiling an entirely new paradigm in the healthcare industry that is changing and improving lives every day. The sky is the limit when it comes to the potential of IoT technology in healthcare, and the journey towards better

© The Author(s), under exclusive license to Springer Nature Switzerland AG 2025
V. Sai et al., *6G Enabled Healthcare Systems*, Synthesis Lectures on Engineering,
Science, and Technology, https://doi.org/10.1007/978-3-031-73849-4_5

Fig. 5.1 Evolution of health
information technology

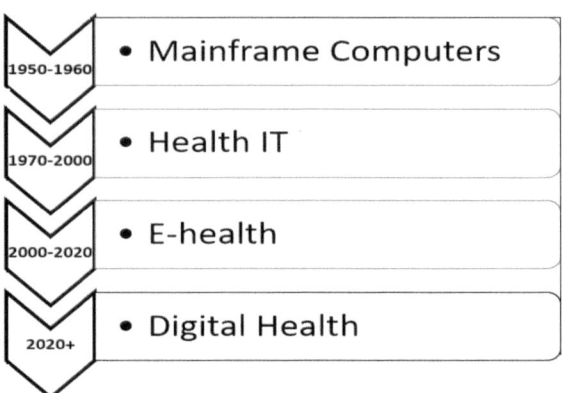

and more efficient care of an individual is an ongoing one that will continue to improve
with time.

5.1   Simple Healthcare Ecosystem

The healthcare industry is rapidly evolving with the introduction of digital technologies.
One such advancement is the development of a basic digital ecosystem that has the potential to promote and improve the health and well-being of patients. Figure 5.2 highlights
the aspects of this ecosystem that contribute to its evolution [3].

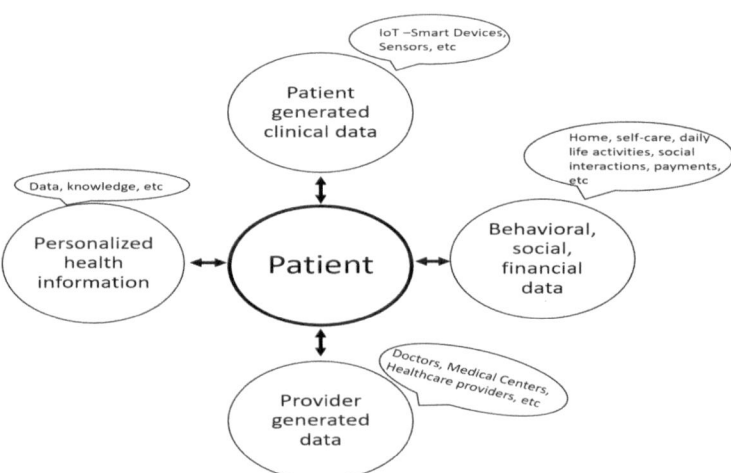

Fig. 5.2 Basic patient's digital healthcare ecosystem [3]

This digital ecosystem includes the interaction of digital technologies with the patient's data. Adaptive technologies are utilized to support and improve the patient's condition continually. It is worth noting that this ecosystem is not limited to only these systems, and there's potential for more advancements in the future.

As healthcare continues to move towards a digital-centric approach, one of the driving forces behind the creation of this digital ecosystem is patient-generated clinical data. The ecosystem tracks and records relevant data using smart devices, sensors, and other tools to collect relevant data. This data collection is not limited to the patient's vital signs but also to environmental conditions, demographics, and lifestyle factors that could influence their health.

As mentioned in Fig. 5.2, healthcare providers generate data that complements the data provided by the patients. These data are collected by physicians, hospitals, caregivers, insurance companies, and other healthcare service providers to aid in the diagnosis and treatment of the patient.

In conclusion, this digital ecosystem is just the beginning of a wide array of technological advancements that will enhance patient care and well-being. With patient-generated clinical data, their behaviour and social patterns, healthcare provider generated data, and the patient's health information being staples of the ecosystem, in the future, there will undoubtedly be more breakthroughs awaiting in the field of health tech.

The patient's involvement in their healthcare is vital to achieving the best health outcomes and medical decisions. One of the key components of patient engagement and involvement is the generation of clinical data. In recent times, technology has been an important contributor to the generation of patient-generated clinical data. Through the use and connection to IoT ecosystem, smart devices and sensors are utilized to collect relevant data that can offer vital insights into the patient's health.

The digital ecosystem tracks and monitors the patient's daily behavioral activities such as habits or actions, remote monitoring, self-care, and financial options/solutions that form an integral part of the patient's ecosystem. Real-time data collection and tracking aids in the holistic understanding of the patient's health, including the environmental conditions, demographics, lifestyle factors, and other aspects that could influence the patient's well-being.

Figure 5.2 emphasizes the importance of personalizing health information for the patient, which enables the patient to take ownership of their health outcomes. Patients receive information related to diagnosis, treatment, medical decisions, and other personalized information unique to their health needs. Access to personal health records (PHR) offers a complete view of the patient's medical history, current treatments, medications, allergies, and other vital information that aids in medical decisions.

Furthermore, healthcare providers generate traditional data that complement the data provided by patients. These data are collected through an in-person support model, where patient data are collected by the doctor, hospital, caregiver, insurance companies, and others involved in patient care. Patient-generated clinical data and provider-generated data work hand in hand to provide a comprehensive understanding of the patient's well-being, leading to better medical decisions and health outcomes.

In summary, generating patient-generated clinical data is necessary for achieving the best patient outcomes. It offers a holistic view of the patient's condition and well-being. With the use of IoT ecosystems, smart devices and sensors, remote monitoring, self-care, and financial options/solutions, data collection has become more efficient, paving the way for the provision of personalized health care services to the patient.

The healthcare industry has undergone a significant transformation in recent years, and one of the driving forces behind this revolution is the implementation of 5P medicine. This medicine focuses on personalized, preventive, predictive, participatory, and precise healthcare, with the aim of revolutionizing the healthcare ecosystem [4]. Figure 5.2 emphasizes that to make the services of the 5P medicine effective, it is necessary to have access to extensive data, including the patient's electronic health records (EHR) in addition to their individual characteristics, treatments, health history, health-related behaviours, medicines, and treatments [5].

Personalized medicine recognizes that every patient is different, and each requires a unique approach to their healthcare treatment. Understanding the patient's genetic makeup, lifestyle and environmental factors is crucial in providing personalized medicine that satisfies the patient's specific needs.

Preventive medicine aims to identify and address health issues before they manifest in the form of symptoms. The approach focuses on preemptive measures to reduce the probability of developing health issues.

Predictive medicine is an essential component of the 5P approach. The purpose of this approach is to use data analytics and AI to predict potential health problems that an individual may develop in the future, leading to preventive actions.

Participatory medicine involves active participation of the patient in their healthcare treatment. It emphasizes the importance of engaging the patient in their health treatment and empowering them to take ownership of their health.

Lastly, the 5P approach focuses on precise medicine, which involves accurate identification of the cause of the disease and the specific treatment that would work for the patient. This approach is made possible by data analysis and precise diagnostic tools.

In conclusion, the 5P medicine offers a holistic approach to healthcare, acknowledging the importance of preventive measures and empowering the patient to take charge of their health. The implementation of the 5P medicine requires access to extensive data, including patient electronic health records (EHR), individual characteristics, treatments, health history, health-related behaviors, medications, and treatments, among others [5]. The 5P approach is the future of healthcare and presents unlimited opportunities for the evolution of healthcare treatments.

Summary 5.1

1. The healthcare industry is transitioning towards a digital-centric approach and utilizing a basic digital ecosystem that involves patient-generated clinical data to monitor health factors such as environmental conditions and lifestyle. Healthcare providers complement this data to provide better diagnosis and treatment.
2. The 5P medicine is revolutionizing the healthcare industry and involves personalized, preventive, predictive, participatory, and precise healthcare. It requires extensive data, including access to electronic health records and individual characteristics to provide unique healthcare treatments for patients.

5.2 Sources of Personal Health Information (PHI)

With the increasing number of IoT devices and numerous sources of information available, personal health information (PHI) collection has shown explosive growth compared to receiving health data using traditional methods. Figure 5.3, which presents various sources of PHI, highlights how information is collected not only from electronic health records (EHR), but also from social media, home care systems, personal health devices, web browsers, and other sources that are not currently present in EHRs [5]. Figure 5.3 continues to highlight that gathering PHI for effective healthcare delivery is no longer enough, and it is equally important to obtain this information when the patient is healthy. This would enable the provision of efficient 5P medical services [5].

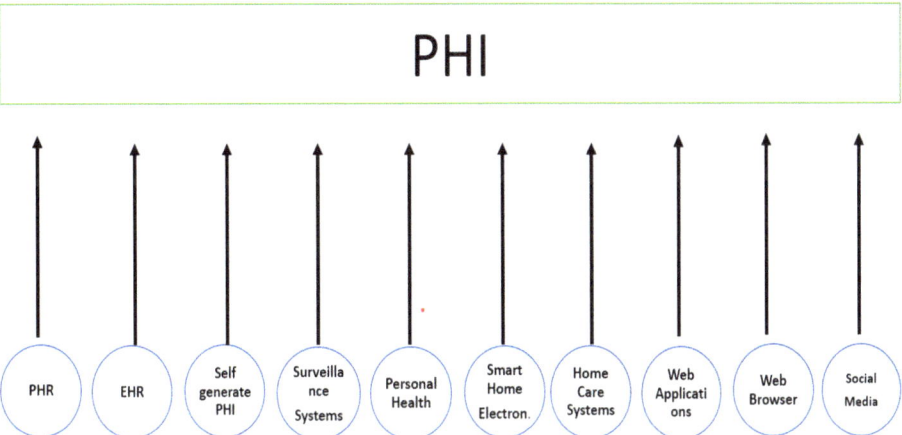

Fig. 5.3 Typical sources of personal health information [5]

One of the primary sources of PHI is EHR. These records contain the patient's medical history, including the treatments they have received, the medications taken, and other essential data necessary for the delivery of healthcare. Other sources of PHI include social media, where patients tend to share their health information, experiences and seek advice from others, and homecare systems that collect data on the patient's activities of daily living or how an individual interacts with their environment.

Personal health devices, such as smartwatches, fitness trackers, and wearable technology, are becoming increasingly popular sources of PHI. These devices track and monitor health-related information, such as the patient's heart rate, calorie intake, sleep patterns, and physical activities, and send this data to healthcare providers in real-time.

Web browsers and applications are other sources of PHI, which could include online searches for diagnosis, treatment options, medications, and home remedies. These sources bring reinforcements to patient-generated clinical data, making PHI sources more diverse, richer in data, and complete for the healthcare provider.

In conclusion, the extensive growth of IoT devices and various sources of information has seen a significant increase in the collection of PHI. Figure 5.3 presents multiple sources of PHI, including social media, home care systems, personal health devices, web browsers, and other sources that would enable effective 5P medical services. EHR remains a primary source of PHI. However, it is necessary to collect PHI not only when the patient is sick but also when they are healthy to provide efficient health care services.

Summary 5.2

1. The collection of personal health information (PHI) has experienced explosive growth due to the increasing number of IoT devices and diverse sources of information, including social media, personal health devices, web browsers, and home care systems, among others. Electronic health records (EHR) continue to remain a primary source of PHI.
2. It is important to collect PHI not only when the patient is sick but also when they are healthy to provide efficient healthcare services using the 5P medicine approach. The diverse sources of PHI reinforce patient-generated clinical data, making PHI sources richer, more diverse, and complete for the healthcare provider.

5.3   IoT Data Flow Lifecycle

The future of healthcare lies in data-driven and personalized care that requires clinicians to have more significant visibility into patient data and important health indicators, while patients actively participate in shared decision making, ensuring patient-centric healthcare services [6]. The healthcare industry has witnessed a gradual integration of IoT technology, among others, to provide better services and efficient results to patients. Figure 5.4, illustrating typical stages of IoT ecosystem data flow, shows how data, particularly health data, flows within an IoT ecosystem [7].

As described in the previous chapter, IoT sensors collect vast amounts of health data from patients that are then processed by the respective medical practitioners to provide accurate diagnoses and high-quality care as efficiently as possible. The IoT sensors communicate with various healthcare providers, which store the collected data in a centralized cloud or a distributed storage system. The collected data is then analyzed, resulting in the generation of accurate reports and precise data visualizations necessary for healthcare delivery services. Through this process, future healthcare services will be based on data insights that would enable healthcare providers to personalize care, promote wellness and disease prevention, and enhance patient participation in shared decision making.

Furthermore, the healthcare sector is investing substantially in machine learning and artificial intelligence (AI) to improve the efficiency and accuracy of data analysis necessary for personalized and data-driven healthcare. These technologies help ensure the correct diagnosis, offer better treatment options, and generate more accurate predictions for wellness and disease prevention.

It is expected that the healthcare industry will continue to witness innovations in IoT technology, health data collection, and data-driven techniques that will continue to enhance personalized healthcare services, improve patient outcomes, and drive better healthcare results for patients. The inclusion of advanced IoT technology with healthcare

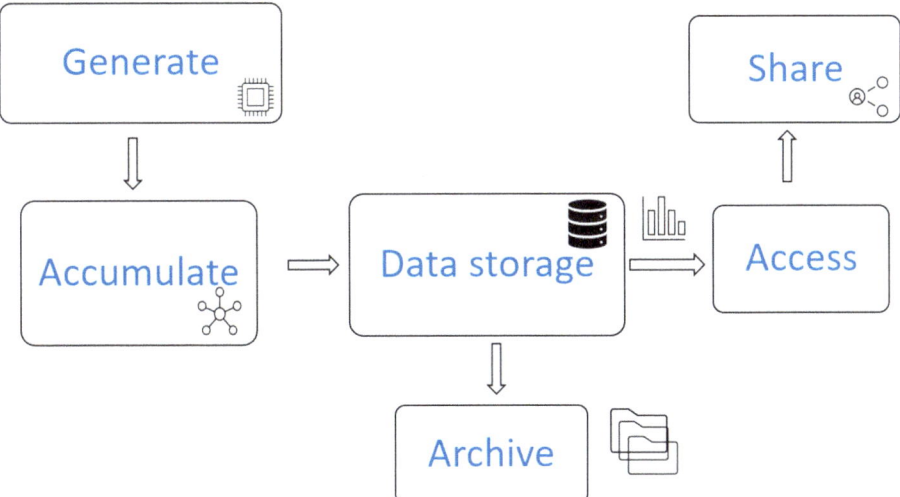

Fig. 5.4 Typical IoT data activity flow [7]

not only revolutionizes the industry, but also offers patients better access to experience personalized care and interventions tailored to their individual needs.

Summary 5.3

1. The future of healthcare lies in personalized and data-driven care that involves active patient participation and engagement in shared decision-making for patient-centric healthcare services with better access to accurate health data.
2. The healthcare industry is gradually integrating IoT technology, machine learning, and artificial intelligence (AI) for better health data collection, data analysis, accurate diagnosis, treatment options, and disease prevention that drive better healthcare outcomes for patients. This revolutionizes the healthcare industry, offering advanced personalized care and interventions tailored to individual needs.

5.4   Use-Case Scenario

In the wake of pandemics, especially COVID-19, IoT devices prove beneficial in monitoring and managing affected patients [8]. Figure 5.5 represents a crucial use case scenario for IoT devices, revealing how technology can help healthcare providers make informed decisions and take the necessary actions to prevent the spread of the disease. The figure illustrates the different environments, including work and home settings, hospitals, emergency vehicles, and unmanned aerial vehicles (UAVs). Through real-time patient data retrieval using IoT devices, the data is aggregated, stored, and processed in real time, helping healthcare providers make real-time decisions to avoid the spread of the pandemic.

In this scenario, the 6G is used to communicate data from the IoT devices to healthcare providers. The 6G communication paradigm offers a high data rate, enabling seamless transmission of large volumes of data and an incredibly low latency rate, which is beneficial when a quick response is necessary to avoid the spread of the pandemic. The added AI capabilities in 6G communication technology work to help healthcare providers make quick decisions to reduce response times, a crucial factor that could save lives.

The IoT devices in Fig. 5.5, including wearables, sensors, and home monitoring systems, provide information about the patient's vital signs, including temperature, respiratory rates, and blood pressure. With these devices, healthcare providers can identify symptoms and predict the development of the disease, making it easier to monitor and treat patients more effectively. The pandemic scenario depicted in the figure also highlights the importance of Big Data analytics, which works alongside the 6G and AI capabilities to ensure accurate data insights.

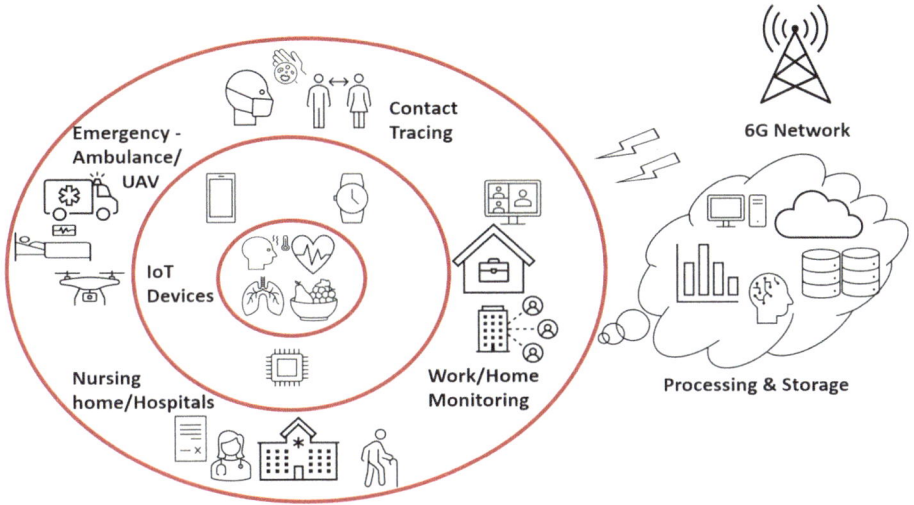

Fig. 5.5 Sample scenarios for IoT devices in healthcare [8, 9]

Overall, the use of IoT devices during pandemics such as COVID-19 would prove beneficial in coordinating and managing patient care efficiently. The innovative 6G communication technology combined with AI capabilities can help healthcare providers make quick decisions to reduce response times and ultimately save lives. The combination of wearables, sensors, and home monitoring systems also offers a new approach to the management of chronic diseases and other health conditions. With continued advancements in IoT technology and related fields, the healthcare industry can continue to improve patient outcomes and enhance overall healthcare service delivery.

Summary 5.4

1. IoT devices are beneficial in monitoring and managing patients affected by pandemics, such as COVID-19.
2. In this scenario, 6G communication technology with AI capabilities helps healthcare providers make quick decisions to prevent the spread of the pandemic.
3. IoT devices, including wearables, sensors, and home monitoring systems, combined with Big Data analytics, enable healthcare providers to monitor and treat patients more effectively, offering a new approach to managing chronic diseases and other health conditions.

5.5   6G Based IoT for Healthcare

Figure 5.6 is an essential reference point that highlights the various applications of 6G-based IoT technologies in healthcare systems [10, 19]. The figure outlines various studies that discuss the practical implementation of 6G wireless communication technology in the healthcare sector, showcasing its potential benefits in different healthcare applications.

One of the applications mentioned in Fig. 5.6 is the application of 6G's URLLC high-quality video streaming for reliable diagnosis in mobile connected ambulances. Cisotto's study presents the use of 6G-based URLLC technology for video streaming that provides reliable diagnosis in ambulances [11]. Another approach to using 6G's potential in healthcare is outlined in Nguyen's study, which describes using 6G edge intelligence for COVID-19 diagnosis [12].

Hadi's study is yet another example of combining 6G with machine learning to develop a healthcare data analytics framework that analyzes patient records and optimizes resource allocation in heterogeneous 6G networks [13]. This framework is useful in managing different healthcare data and using available resources more efficiently.

Mucchi's study discusses an advanced healthcare data communication paradigm that uses 6G-based mURLLC and various IoT devices [14]. This system enables secure and fast data transfer between IoT devices, ensuring reliable communication.

In another study by Sodhro, a novel solution for mobility management in 6G-based health networks is presented, where machine learning algorithms are employed to manage patients' movement efficiently [15].

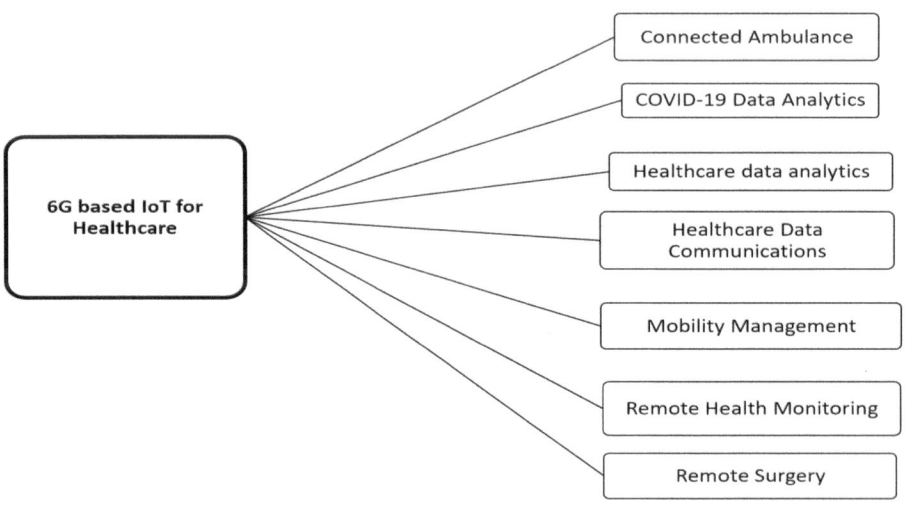

Fig. 5.6 Different 6G-IoT applications [10]

One of the use cases of remote health monitoring, mentioned by Nayak, pertains to utilizing mURLLC and THz-based communication features of 6G for enabling remote doctors to monitor patients [19]. This application has been particularly helpful during the COVID-19 pandemic, where remote patient monitoring is essential to protect the healthcare workers and patients.

Lastly, Fig. 5.6 also references the integration of unmanned aerial vehicles (UAVs) and blockchain. Gupta's study highlights how the combination of these two technologies can enhance the surgery processes remotely, promoting greater accessibility to healthcare facilities for everyone [17].

These studies, along with numerous others, showcase the potential of 6G integrated with IoT technology in healthcare applications [16, 18]. The various applications explored in Fig. 5.6 provide useful insights into the benefits of using 6G wireless communication technology in healthcare systems.

Summary 5.5

1. Fig. 5.6 highlights the various applications of 6G-based IoT technologies in healthcare systems, showcasing its potential benefits in different healthcare applications.
2. The different studies discussed in Fig. 5.6 outline the use of 6G's URLLC technology for reliable diagnosis in mobile connected ambulances, COVID-19 diagnosis, healthcare data analytics, advanced healthcare data communication paradigm using 6G-based mURLLC and IoT, mobility management using machine learning algorithms, and remote health monitoring using mURLLC and THz-based communication features of 6G.
3. The integration of unmanned aerial vehicles (UAVs) and blockchain can enhance the surgery processes remotely, promoting greater accessibility to healthcare facilities for everyone.

5.6 Challenges in 6G Based IoT for Healthcare

6G technologies are driving the advancement of IoT services and technology, offering numerous benefits, such as low latency, high sustainability, efficient scalability, and improved reliability [19]. However, these benefits come at a cost, especially in the context of healthcare applications where 6G poses unique challenges [8, 10, 18]. For example, the healthcare industry requires high data precision, speed, and reliability for efficient diagnosis, treatment, and monitoring. Given that 6G networks are still in the development phase and are set to be rolled out in the coming years, there is a need for careful planning and deployment of these new technologies.

One of the primary challenges facing healthcare providers in implementing 6G-based IoT applications is the cost of upgrading the existing infrastructure to support these emerging technologies. The process of upgrading infrastructure can be time-consuming and requires considerable investment. Furthermore, the implementation of 6G-based technologies involves complex and advanced technologies that require a skilled workforce that can effectively manage and oversee the new infrastructure.

Another challenge for healthcare stakeholders in implementing 6G-based IoT applications is ensuring data security and privacy [10]. Healthcare data is highly sensitive and confidential, and any breach could have severe consequences. The need for secure data transfer becomes even more essential when implementing IoT applications in the healthcare system, where data is transferred between different devices and systems [8].

Furthermore, in healthcare, latency is a significant concern, especially when dealing with monitoring and treatment applications. 6G promises to have low latency, but its actual performance in the healthcare ecosystem needs to be tested extensively. There is also the issue of 6G-enabled devices' interoperability, which must exist for seamless communication across devices.

In the end, despite the numerous benefits of 6G-based IoT applications in healthcare, such as improved connectivity, better diagnoses, and treatment, challenges cannot be ignored. The cost of upgrading infrastructure, ensuring security and privacy, addressing latency concerns, and interoperability issues are critical factors that must be considered before implementing these technologies into the healthcare ecosystem. Proper planning and execution of implementation can help healthcare organizations leverage the benefits offered by 6G technologies while appropriately mitigating the challenges and limitations for better outcomes.

The implementation of 6G-based IoT technology in various sectors including healthcare comes with its fair share of challenges, one of which is ensuring adequate security and privacy protocols are in place. As healthcare initiatives are now dependent on the efficient collection, transfer, storage, and processing of large volumes of patient data, it becomes crucial to prevent any unauthorized access to sensitive information, keeping it out of the wrong hands.

To achieve optimal security, sensitive patient data involved in 6G-based IoT must be protected from potential cyber attacks during the data collection and transfer phase to IoT devices. The system's wireless network used to transport this data should also be secure and protected against any cyber-attacks as they pose serious threats to any healthcare facility. Subsequently, the collected data must also be processed and stored on secure servers equipped with the latest encryption technology and access control.

In this regard, measures such as passwords, biometric data authentication, data encryption, and firewalls may be utilized to fortify access control, only allowing authorized access to relevant parties. The entire system should be designed with layered security from each IoT device in the network to the data transmission lines.

Furthermore, strict protocols must be put in place for the handling of sensitive patient information by healthcare professionals who will have access to these data. This step requires first that the staff are trained in cybersecurity, and also adhere to HIPAA (Health Insurance Portability and Accountability Act) privacy standards, especially when it concerns sensitive medical or personal information for specific individuals.

On the whole, the challenges posed by the security and privacy of the 6G-based IoT are considerable. The high stakes of the healthcare industry require that continued support for the development of the latest solutions to secure IoT in healthcare will continue. However, with strategies such as access control and encryption, it is possible to safeguard sensitive patient data with the latest technologies, while ensuring that only authorized personnel can work with the information.

In 6G-based IoT systems, ensuring resource efficiency is of great significance, particularly in ensuring that all continuous monitoring, processing, storage, communication, and network operations are effectively managed [10, 18]. Resource efficiency entails the proper handling of several hardware constraints related to power usage, computations, and storage. Resource efficiency is critical in ensuring the optimization of the entire 6G IoT system while enhancing its reliability and scalability.

When dealing with healthcare data analytics, resource efficiency poses a considerable challenge, especially when edge devices are used to provide quality diagnosis to patients. Healthcare data analytics can require high computing power coupled with the necessary storage and low-power-based systems to cater to the specific needs of patients. It is important to ensure that the computational and storage resources available are of appropriate capacity to cater to the needs of the entire system in question, including the edge devices.

Resource optimization also requires efficiency in network operations and communication throughout the entire system. This translates to establishing and maintaining quality service levels in bandwidth utilization, message and data transmission reliability, network coverage, and latency. With this in mind, our focus should be on incorporating the latest technological advancements in AI, machine learning, and edge computing to provide an optimal resource-efficient framework that would improve overall system operation.

Furthermore, power consumption is a critical factor when dealing with resource efficiency in 6G IoT systems. With edge devices becoming more commonly used in healthcare, it becomes a challenge to maintain the necessary power levels required for the devices while optimizing overall performance. Therefore, necessary measures must be implemented to ensure that the power consumed through devices remains minimized by using energy-efficient components to enhance power management.

In essence, resource efficiency is essential to successfully optimize the performance of 6G-based IoT systems. Effective resource handling of power usage, computation, and storage are critical factors that need to be considered. The necessary hardware and communication systems must receive appropriate implementation to ensure network operations and communication efficiency. Incorporation of modern technological advancements such as AI, real-time processing, and edge computing is necessary to produce resource-efficient frameworks catered to specific needs of the healthcare sector.

Regulation and standardized compliance constitute a cardinal aspect of the operationalization of 6G-IoT based systems [8, 10, 18]. This guarantees or ensures inter-operability among all network providers, vendors, services, autonomous devices, and aggregated data, thus enhancing service quality and customer satisfaction. The achievement of this mandate is necessary to promote optimum performance for the systems involved within the 6G-IoT ecosystem.

Incompatibility challenges exist between IoT devices concerning interfaces and protocols, significantly hindering effective communication, data aggregation, functional operation, and productivity. When operationalizing healthcare-oriented IoT devices, standardization, and regulation of the underlying technology become even more vital towards the effective management of health outcomes. Using a standardized set of measures ensures that the different networks harmonize and integrate seamlessly, resulting from the 6G-IoT ecosystem's integrated framework.

Furthermore, having well-established regulations is an effective approach to address the potential security threats emanating from 6G-IoT solutions, particularly in the healthcare industry. The health sector is highly vulnerable to data breaches hence the need for thorough regulatory and governance measures to mitigate these risks. Data breaches can cause not only operational, but also commodity damage, leading to prolonged downtime and severe financial losses for healthcare providers.

Conversely, the compliant application of regulations ensures that the devices' operations are aligned towards assigned norms, enabling the healthcare information's adequate protection. It also advocates for appropriate offline and online backups of all stored information, ensuring business continuity and data restoration during any eventualities.

In conclusion, standardization, and compliance to regulatory policies is of utmost importance in integrating and optimizing the performance of 6G-IoT systems. In healthcare-based IoT devices, the standardization aspect is essential to ensure successful operations, mitigate security risks, and ensure seamless integration across various networks. The application of regulations enforces strict adherence to well-established rules, protecting healthcare information and promoting business continuity. The provision of safety measures is a further guarantee or layer of disincentive to cyberattacks and data breaches, ultimately putting the healthcare industry at ease.

Summary 5.6

1. Healthcare faces unique challenges in implementing 6G-based IoT applications due to high data accuracy, speed, and reliability requirements.
2. Ensuring security and privacy protocols are in place for 6G-based IoT systems in the healthcare industry is crucial.
3. Standardization and compliance to regulatory policies are important in optimizing performance and ensuring the protection of healthcare information in 6G-IoT systems.

References

1. A. Snowdon, *Digital Health: A Framework For Healthcare Transformation, White Pape*r. (Healthcare Information and Management Systems Society, 2022)
2. FDA. What is Digital Health? (2020). https://www.fda.gov/medical-devices/digital-health-center-excellence/what-digital-health. Accessed 1 February 2023
3. A. Marinescu, C. -M. Rotaru, D. Nicolau and P. Krawiec, Challenges and perspectives for the development of a future ecosystem for elderly within pandemic, in *2021 23rd International Conference on Control Systems and Computer Science (CSCS)* (Bucharest, Romania, 2021), pp. 501–508. https://doi.org/10.1109/CSCS52396.2021.00088
4. G. Pravettoni, S. Triberti, *P5 eHealth: An Agenda for the Health Technologies of the Future* (Springer Open, Berlin, 2020), pp. 3–17
5. P. Ruotsalainen, B. Blobel, Transformed health ecosystems—challenges for security, privacy, and trust. Front. Med. **9** (2022)
6. S.J. Park, M. Subramaniyam, S.E. Kim, S. Hong, J.H. Lee, C.M. Jo, Y. Seo, Development of the elderly healthcare monitoring system with IoT, in *Advances in Human Factors and Ergonomics in Healthcare.* (Springer, Cham, 2017), pp.309–315
7. A. Abugabah, N. Nizamuddin, Smart healthcare ecosystem for elderly patient care, in *2020 43rd International Convention on Information, Communication and Electronic Technology (MIPRO)* (Opatija, Croatia, 2020), pp. 365–370.
8. A. Koren, R. Prasad, IoT health data in electronic health records (EHR): security and privacy issues in era of 6G. J. ICT Stand. **10** (2022)
9. S.N. Nova, M.S. Rahman, C. Chakraborty, Patient's health surveillance model using IoT and 6G technology, in *Green Technological Innovation for Sustainable Smart Societies.* ed. by C. Chakraborty (Springer, Cham, Switzerland, 2021), pp.191–209
10. D.C. Nguyen, M. Ding, P.N. Pathirana, A. Seneviratne, J. Li, D. Niyato, O. Dobre, H.V. Poor, 6G internet of things: a comprehensive survey. IEEE Internet Things J. **9**(1), 359–383 (2022)
11. G. Cisotto, E. Casarin, S. Tomasin, Requirements and enablers of advanced healthcare services over future cellular systems. IEEE Commun. Mag.Commun. Mag. **58**(3), 76–81 (2020)
12. D. Nguyen, M. Ding, P.N. Pathirana, A. Seneviratne, Blockchain and AI-based solutions to combat coronavirus (COVID-19)-like epidemics: a survey. IEEE Access (2021)

13. M.S. Hadi, A.Q. Lawey, T.E.H. El-Gorashi, J.M.H. Elmirghani, Patient-centric HetNets powered by machine learning and big data analytics for 6G networks. IEEE Access. **8**, 85 639–85 655 (2020)

14. L. Mucchi, S. Jayousi, S. Caputo, E. Paoletti, P. Zoppi, S. Geli, P. Dioniso, How 6G technology can change the future wireless healthcare, in *Proceedings of the 2020 2nd 6G Wireless Summit (6G SUMMIT)* (Levi, Finland, 2020), pp. 1–6

15. A.H. Sodhro, N. Zahid, L. Wang, S. Pirbhulal, Y.O. Ouzrout, A. Sekhari, A.V. Lira Neto, A.R.L. De Macedo, V.H.C. De Albuquerque, Towards ML-based energy-efficient mechanism for 6G enabled industrial network in box systems. IEEE Trans. Ind. Inform. 1–1 (2020)

16. H. Habibzadeh, K. Dinesh, O.R. Shishvan, A. Boggio-Dandry, G. Sharma, T. Soyata, A survey of healthcare internet of things (HIoT): a clinical perspective. IEEE Internet Things J. **7**(1), 53–71 (2020)

17. R. Gupta, A. Shukla, S. Tanwar, BATS: a blockchain and aiempowered drone-assisted telesurgery system towards 6G. IEEE Trans. Netw. Sci. Eng. 1–1 (2020)

18. S.M.R. Islam, D. Kwak, M.H. Kabir, M. Hossain, K.-S. Kwak, The internet of things for health care: a comprehensive survey. IEEE Access **3**, 678–708 (2015)

19. S. Nayak, R. Patgiri, 6G communication technology: a vision on intelligent healthcare. IEEE Internet Things J. **7**(10), 9227–9234 (2020)

One of the key features of 6G compared to the existing 5th generation of cellular networks is support for high-speed (traveling speed typically measured in miles or kilometers per hour rather than data transmission speed in number of bits per second) vehicular communication in the magnitude of Gbps for downloading requirements even when a user is traveling in a fast moving vehicle. The rapid evolution of these standards advancing from High Speed Packet Access (HSPA) to Long Term Evolution (LTE) and more recently 5G have made telemedicine systems more reliable with a more comprehensive suite of healthcare and medical services made available to both patients and medical professionals. From the network operator's point of view, adoption of a new set of standards when moving into 6G (and associated equipment) heavily rely on being able to convince its subscribers that the new technology is better. Ultimately, the cost of adoption will be met by consumers and patients.

A comprehensive coverage on the technical details of 6G mobile communications would require hundreds of pages in more than one volume for each standard. While this volume deals with various aspects of wireless communications, this chapter is aimed at providing a brief introduction to 6G enabled healthcare systems [3]. We shall conclude the chapter by a discussion of the future directions and 6G in supporting a diverse range of telehealth services.

6.1   An Overview of Cellular Wireless Communication Leading to 6G

The popularity of wireless communication technology has increased spectacularly over the past decade sparked by the cellular telephony market growth. This is followed by the technological advancements in mobile computing device designs and feature enhancements of mobile phones. The first generation of cellular communication appeared around 1980 with each new generation appeared roughly a decade later. We have now observed some four decades of evolution from the first generation analog system to the current mainstream 5G through to the development of 6G.

Perhaps the most significant milestone for comprehensive and pervasive healthcare systems has commenced through the transition from 3 to 4G [15]. The 4G standard was developed by the Third Generation Partnership Project (3GPP) which begun in November 2004, built upon the Universal Mobile Telephone System (UMTS) and is also known as Wideband Code Division Multiple Access (W-CDMA). One of the key approaches is the use of MIMO (multiple-input and multiple-output) that utilizes smart antenna technology with multiple antennas at both the transmitter and receiver ends enabling spatial processing for enhanced spectral efficiency [19]. One of the main design objectives of 5G is substantially reduced latency [36], as well as increased availability and network capacity. These are all essential attributes for a reliable network to support real-time healthcare services.

Digital cellular data service commenced with GSM (Global System for Mobile communications) in the 1990s that operate in the frequency band of around 880–960 MHz, the 2.5 Generation system supports data rate of up to 9.6 Kbps. This rate was then increased in 2 stages. The next data standard, High Speed Circuit Switched Data (HSCSD), utilizes multiple TDMA to assign time slots per connection to support a maximum data rate of 38.4 kbps by using 4 channels simultaneously. Followed by General Packet Radio Service (GPRS) that reuses existing GSM radio channels to provide up to 8 channels simultaneously, supporting a maximum data rate of 140 kbps. It is a packet oriented service that will more efficiently transport bursty data. Wireless Cable, also known as MMDS (Multichannel Multipoint Distribution Service), that in turn supports the delivery of primitive healthcare data to subscribers. Using the frequency band of 2.6 GHz, it also included channels for Instructional Television Fixed Service (ITFS). Channelization allows interactive data services to be supported by MMDS.

Ever since the invention of cellular communication, the successful rollout of the next generation is determined by the ability of a network operator and the equipment manufacturers to deliver products and services that complies with the applicable standards and that the operator can introduce enhancements smoothly without any interruption to services. Since the consumer mobile device is the primary connection that links a subscriber to the network operator, one driving force in moving subscribers to the next generation would be incentives for upgrading the mobile device. In many developing countries, especially those with vast geographical coverage, there is an insatiable demand for wireless communications as rapid growth prospers rural areas where legacy plain old telephone service (POTS) infrastructure have not been well established.

Moving to the 6G era, far higher capacity that can potentially be 100 times that of 5G and much lower latency that is substantially shorter than one millisecond. Coupled with location awareness, it is possible that operations such as natural disaster recovery can be supported by 6G networks with mobile edge computational capabilities to be built into 6G networks. Such properties make 6G particularly suited for supporting highly-customized health anomaly detection applications.

6.1.1 Healthcare Systems Evolving from 3 to 6G

The process of moving from 3G all the way to 6G involved many industry sectors from network operators to healthcare service providers and telecommunications equipment as well as medical device manufacturers, collective efforts transform numerous technical specification documents into physical products that provide services to consumers and patients who will voluntarily pay money for their mobile devices and the recurring usage. The demand for mobile broadband data access has been increased substantially over the past decade since mobile phone users are no longer limited to using their phones merely for voice communications. The need for a more advanced cellular communication system than 3G that has been widely used for over a decade leads to the development of self-organizing networks (SON) when moving into 4G, with the advantages of self-configuration, self-optimization and self-healing for enhanced efficiency and reliability. LTE SON is an approach of cognitive radio aspects in cellular communication development [12]. Information about the power and spectrum allocation of neighboring cells is necessary in order to maintain the femtocell coverage and minimize cell-to-cell interference. A femtocell unit integrates the functions of a typical base station (eNodeB), similar to that of a Wi-Fi access point and also includes all the core network elements. In 3GPP terms, an LTE femtocell is referred to as Home Node B for HSPA (HNB) and Home eNode B for LTE (HeNB). Femtocells must be able to search the radio channel and evaluate which network resources are available in order to avoid cross-layer and co-layer interference [2]. The core network that only handles data in packet form has the advantages of reduced latency and higher throughput and cost efficiency.

The network performance became adequate for supporting healthcare systems and services when HSPA networks became mature [11]. A wide range of consumer healthcare and medical devices can be linked to different IoT-enabled devices to provide services from general health assessment to mission-critical rescue missions. From the 2G General Radio Packet Service (GRPS) era in the 1990s, consumers expect to be able to retrieve data through their mobile phones. Since then, incentives for switching from one generation to the next are primarily driven by a range of feature enhancement services such as more secure mobile commerce, mobile payments and mobile wallets, feature-rich interactive learning, smooth streaming music and video; as well as more pervasive and reliable mobile health services. There are different technological requirements for each of these services ranging from information security to network stability and connectivity compliance.

Average revenue per user (ARPU), calculated from the total revenue divided by the number of subscribers, i.e. the revenue generated by one consumer subscription per month or annum, is a number that a network operator strives to maximize for profitability. To increase this number, the network needs highly monthly caps and download data throughput in order to generate higher data ARPU, gaining network capacity in terms of both the number of users supported and data throughput. At the same time, reduction to the data cost structure while experiencing a vast increase in mobile data traffic with ever-increasing multimedia services, particularly important with unlimited data or flat-rate subscription plans [25]. All these push for a brand new network in situations where existing network is saturated and that scalability incurs hefty upgrades.

6.1.2   The 4G War Competing Standards: LTE Versus Mobile WiMAX

The evolution of 4G mobile communications entailed both Long Term Evolution (LTE) and Mobile WiMAX (Worldwide Interoperability for Microwave Access) due to historical reasons. Following the two parallel paths made standardization for healthcare service support particularly challenging. LTE, based on the 3G communication standard, was built upon the original Global System for Mobile (GSM) communication standard developed by the European Telecommunications Standards Institute (ETSI) Technical Specification (TS), followed by the subsequent development of 3GPP TS. Whereas WiMAX is evolved from the IEEE 802.16 standard (please refer to 6.2.1 for more details) originally designed for fixed wireless communications. While both LTE and mobile WiMAX are designed to provide high-speed ubiquitous broadband wireless access and are IP networks based on OFDM (orthogonal frequency-division multiplexing) technology, both LTE and mobile WiMAX use similar modulation schemes hence they have similar spectral efficiency. To the network operator, LTE provides a more natural upgrade for their existing Groupe Spécial Mobile (GSM) or High-Speed Downlink Packet Access (HSDPA) networks. As with other IEEE 802.X family of standards, WiMAX is built upon an open standard that involved a large community of engineers during its development stage. Such openness implies WiMAX equipment is standard and that anyone is permitted to manufacture WiMAX-compliant equipment. In contrast, developers determining the LTE standard through the 3GPP are formed by a group of network operators and equipment manufacturers who have been trading and using proprietary equipment without open competition.

The WiMAX standard, on the other hand, is evolved from the Local Multipoint Distribution Service (LMDS) platform. It was originally intended to serve as a point-to-multipoint (PMP) last-mile solution [4]. LMDS initially rolled out with carrier frequencies in the 10–40 GHz range with frequency selection primarily depending on local licensing as well as geographical region. Generally, topical areas with persistent heavy rainfall tend to utilize carriers of lower frequencies [9]. The radio range is limited to 3 km due to rain fade attenuation. With the network architecture shown in Fig. 6.1, it was primarily designed as a fixed broadband wireless access (BWA) network. In this network, there are three major components, namely switching where connection to the outside world is provided, transport where connection to its subscribers are supported, and access where subscribers have their user devices located and are served by the cellular communication system. Each of these components has a critical role to play in the successful deployment of the overall system and they can have a significant impact on operational reliability and performance. One of the key elements of supporting mobility is session continuity for real-time applications including both voice and data communication. So, one fundamental difference between supporting fixed and mobile subscribers is whether connection is provided by one or multiple basestations. This leads to the requirement of "handover" operations.

The licensed frequency band can differ significantly in different countries. For example, in the USA LTE is implemented in the 700 MHz band whereas mobile WiMAX is implemented in the 2600 MHz band hence LTE has a significant advantage in coverage range over mobile WiMAX as shown in Fig. 6.2. Although 700 MHz offers much wider coverage, MIMO antenna would be much more difficult to implement in mobile phones operating at 700 MHz due to small geometries in relation to the longer wavelength of the carrier wave.

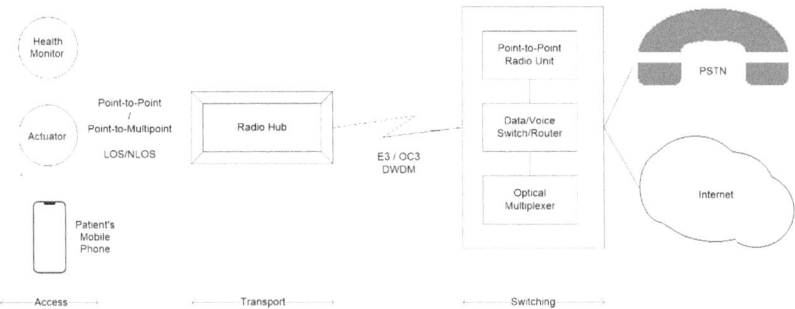

Fig. 6.1 Simple healthcare system architecture

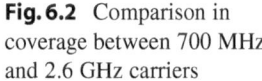

Fig. 6.2 Comparison in
coverage between 700 MHz
and 2.6 GHz carriers

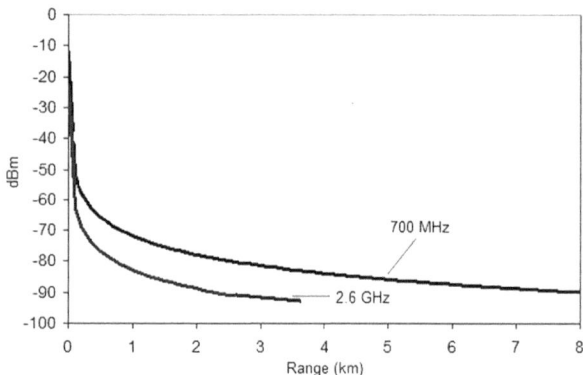

6.1.3 Technological Advances in Cellular Networks for Supporting Healthcare Services

Technological advances bring a range of advantages to all parties involved. To the consumers they expect feature and performance enhancements [5], the network operator drives down the cost of supporting its subscribers in terms of data volume and the number of subscribers through efficient allocation of resources and optimization by network scalability [10]. The equipment manufacturer has the sole objective of selling as many of its products as possible. Attracting as many subscribers as possible benefits both network operator and equipment manufacturer, while driving down the cost per a unit of data (e.g. MB) and provides enhanced coverage is what consumers expect from technological advancements.

LTE supports a 50 Mbps upload rate and 100 Mbps download rate for every 20 MHz of spectrum, which is quite adequate for supporting many health monitoring applications like stroke detection [27] and dietary recommendation [29]. The download rate can be boosted to 320 Mbps with a 4×4 MIMO multiple antenna configurations. LTE provides scalable bandwidth from 1.4 to 20 MHz in both the uplink and the downlink.

LTE Advanced (LTE-A) is a mobile communication standard that enhances the capability of the original 4G LTE standard [21]. Perhaps the most important enhancement from LTE is that LTE-A supports a theoretical peak data rate of 1 Gbps when the subscriber moves at a low vehicular speed. Also, LTE-A provides a generic framework for the support of the interworking between different types of legacy cellular networks [31]. LTE-A introduces a number of features such as Coordinated Multipoint (CoMP) for situations concerning boundaries of multiple cells, Carrier Aggregation for situations where a large chunk of spectrum is split into different sub-channels, for example, five 20 MHz can be aggregated for a total of 100 MHz bandwidth. To combat the effect of multipath using MIMO diversity [33], Multimedia Broadcast and Multicast Service MBMS is configured in such a way that basestations use the same transmission frequency for improved handover.

After competing standards in the earlier 3G (UMTS vs. CDMA2000) and 4G (LTE vs. WiMAX), more recent 5G and 6G are developed with one standard. Cross-platform interoperability should be made much easier with a unified standard.

6.1.4 Green Communications

Mobile communications has become an essential part of life for people from schoolchildren to senior citizens across the world. The substantial rise in both the number of users and data throughput contribute to an increase in energy consumption. To a network operator, the cost of powering its basestations accounts for a substantial share of its operating expenses [16]. The idea of green communication is to reduce the energy consumption while optimizing network capacity in terms of both supporting a larger number of subscribers and the amount of data traffic. In early 2012, the IEEE Communications Society has formally established the Technical Subcommittee on Green Communications and Computing (TSCGCC) that reinstates the importance of promoting global research and development of green technology in telecommunications. For over a decade, substantial work has done to reduce power consumption both at the base station end and the end user's device.

The drive for energy-efficient green communications involves far more than educate consumers to unplug their mobile phones when the battery is fully charged and to power off their devices when not in use. Green communications require cross layer design in communications stack and implementation related to components, network operation and topology [18]. The complexity of modeling and sensitivity analysis for the computation of any change made to any layer of the protocol stack can have significant impact on many basestations across a metropolitan area. Deployment of green communications can yield cost reduction and damage to the environment. Cognitive radios can sense the operating environment and self-adapt to changing conditions. This can easily be used to monitor the channel condition so that dynamic power control can be deployed for improved energy efficiency. Transmission power can be adaptively scaled in response to changing traffic requirement and radio link conditions.

In the cellular communication network, the radio and power amplifier of a base station contribute towards a significant portion of the power consumption [23]. While static power consumption can be reduced by controlling timeslot and channel shutdown, the relative position between an antenna and its radio can be optimized by reducing the length of feeder cables thereby reducing loss. Another important attribute is reduction of the number of site for a given coverage area, which can usually be accomplished with topology specific design and high resolution planning, multiple small base stations in a distributed base station architecture replacing one macro base station would yield a significant power reduction.

6.1.5   The ISM Band

The Industrial Scientific Medical (ISM) band is a portion of the spectrum allocated for a number of applications as its name implies. This was originally assigned to the 4G circa 2.3 ~ 2.6 GHz WiMAX/LTE/TD-LTE bands for mobile communication service and allowed operations of 4G and other ISM within the same device or in close proximity to one another. Although the range of frequencies defined by the ITU spanning across 6 MHz up to as high as 245 GHz, ISM frequencies around 2 GHz has been designated for wireless communications throughout the world [22]. Although network operators are licensed for specific frequencies in many countries, regulatory authorities may allocate adjacent parts of the radio spectrum for unlicensed unlicensed operations meaning that communications can still be subject to interference from other devices. Since the ISM band is also shared by various medical devices and applications, such as radio-frequency process and electrically induced heating for medical treatment [14]; these applications for electromagnetic tissue heating in therapeutic medicine can cause severe electromagnetic interference to communication systems operating in the ISM band.

Electrosurgery utilizes high frequency A.C. electrical current as a cutting modality in order to cauterize small blood vessels, controlled by the frequency and power, to stop bleeding [1]. Such process of high frequency A.C. cutting implies that physical therapy can induce a significant amount of interference that impairs wireless communications. The degree of heat and depth of penetration is determined by the absorptive and resistance properties of the tissues that the generated waves, heat generated by diathermy devices can therefore cause a rapid change to wave generation thereby causing a sudden increase in EMI.

Summary 6.1 An Overview of Cellular Wireless Communication Leading to 6G

1. While 6G networks are promised to offer substantial enhancements over 5G networks currently deployed throughout the world, many physical limitations and operational constraints remain unchanged from earlier generations of cellular networks and systems.
2. Substantial enhancements adequate for supporting mission-critical healthcare services commenced from 4G circa 2012 when networks evolved to become mature enough in terms of performance and reliability. Each iteration from 4G to 5G and 6G takes a significant step forward in providing more types of healthcare services to a larger number of patients with more comprehensive coverage (both in terms of healthcare and geographical coverage). A common standard turns out to be vitally important for supporting a diverse range of healthcare services due to issues such as interoperability and compability.

3. The development in green communications is becoming increasingly important as wearable and implantable devices are made smaller. Cognitive radios have tremendous potentials in making devices more energy efficient, consequently lighter and smaller.

6.2 Mobile Versus Fixed 6G Networks

As with virtually any communication network, the primary objective of either a mobile or fixed network is to deliver the data to its intended recipient in the fastest and most reliable way possible. For a mobile device user to communicate with another person or device, it must be served by some kind of fixed system that gives a wireless access point in order to provide the necessary connectivity to the rest of the networked world.

Dating back to the 4G era, one of the important design considerations was for high speed vehicular communications, where high speed specifically refers to the travelling speed of the user or one's vehicle. It is theoretically capable of providing support for mobility maintained as high as 350 km/h [28], which would be more than adequate for supporting smart ambulances and drones. The design and implementation of 6G networks for supporting smart ambulances in a smart city environment will be discussed in Sect. 7.4.

6.2.1 An Overview of the IEEE 802.16 Standard for Fixed Wireless Communications

While this may be outdated from the design and implementation of 6G systems for telemedicine, this is still highly relevant to fixed communications such as between buildings and within a smart city infrastructure. The IEEE 802.16 standard is of particular historical significant in the development of smart city infrastructures being the standards for broadband wireless metropolitan area networks. From LMDS to WiMAX, where both belong to the IEEE 802.16 standard, WiMAX spans over a much wider range of frequency bands covering 2–11 GHz unlicensed and 10–66 GHz licensed. WiMAX has a significantly extended theoretical maximum range of 50 km with a direct line-of-sight (LOS). The development of WiMAX entails two different branches as illustrated in Fig. 6.3, namely fixed WiMAX is based on the IEEE 802.16–2004 standard and is optimized for fixed receivers like broadband modems and routers; and mobile WiMAX is based on the IEEE 802.16e standard and targeted primarily for mobile devices such as smartphones and tablet computers. The IEEE 802.16 Standard for Air Interface for Broadband Wireless Access Systems with Advanced Air Interface uses scalable OFDMA. Some of the key enhancements are listed in Table 6.1. Certain attributes are required for mobile WiMAX when LOS is not always maintained between the basestation and the user, these include channel bandwidth scalability, frequency reuse, handoff and flexible power management.

The network performance in terms of both coverage and data throughput of WiMAX system is determined by many factors, as shown in Table 6.2.

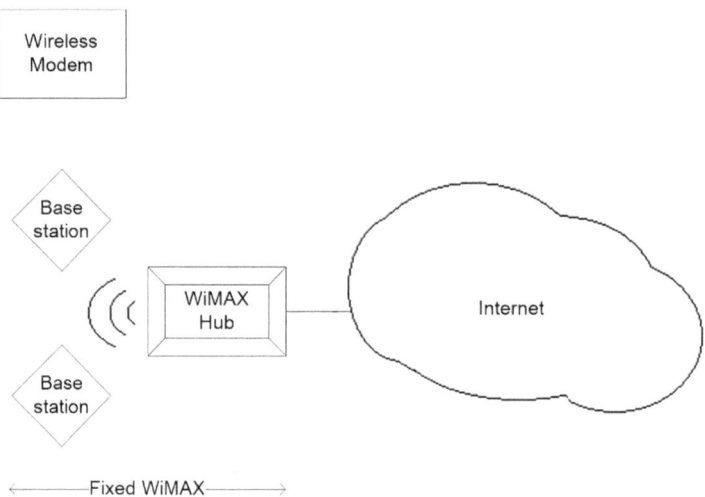

Fig. 6.3 Fixed network architecture

Table 6.1 A comparison between the IEEE 802.16e and m standards

Parameter	Unit	802.16e	802.16 m
Carrier frequency	GHz	2.4–3.5	6–11
Aggregate data rate	Mbps	63	100
Maximum range	km	10	30
MIMO support	No. of streams	4	4–8
Handover interruption time	ms	40	30

Table 6.2 Common factors that affect fixed network performance

Parameter	Deployment options
Frequency band	2–11 GHz unlicenced or 10–66 GHz licenced
Modulation scheme	M-ary QAM: 4 (QPSK), 16 or 64
Line-of-Sight	LOS or NLOS
Number of users	How many users in the basestation sector
Transmission power	Depends on type of device

6.2.2   Trade-Off Between Mobility and Capacity

The capacity of a given basestation, in terms of both the data throughput and number of subscribers within a coverage area of variable size, depends on many factors. We begin by taking a brief look at the effect of using different order of modulation, using a simple example from QPSK to 64 QAM. The trade-off between coverage area and data throughput can be understood from Fig. 6.4 and Table 6.3, respectively. From Fig. 6.4, we can see that QPSK provides a much broader coverage than using higher order modulation schemes; whereas Table 6.3 shows us that we have an improved spectral utilization efficiency (SUE), i.e. more b/s/Hz, when the order of modulation increases. Higher order modulations are more bandwidth efficient, but more susceptible to noise and demands more complex receiver structure since constellation points are more closely packed together. Taking QPSK as an example, decision at the receiver between symbol '00' and symbol '01' is less easy to resolve than a binary decision between '0' and '1'. 16-QAM (4 bits/modulation symbol) and 64-QAM (6 bits/modulation symbol) modulations have symbols even more closely packed. In summary, an increase in data symbol modulation is more spectrum efficient but also less robust.

The theoretical maximum capacity C_{max} that corresponds to the number of correctly transmitted bits per second of the channel, based on Shannon-Hartley Theorem, with channel bandwidth B and additive white Gaussian noise [30]. Given that:

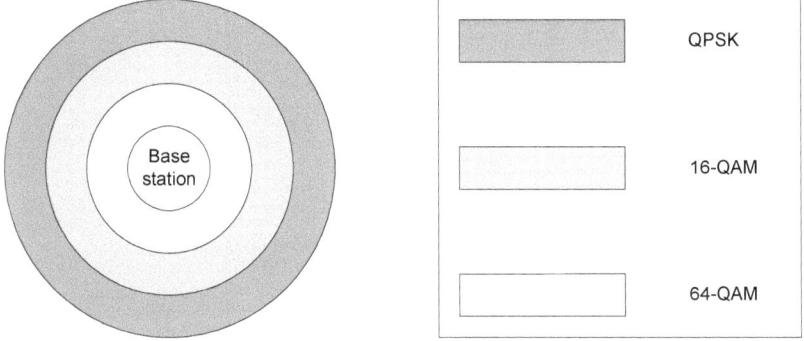

Fig. 6.4   A 2-D representation of coverage with different modulation schemes

Table 6.3   Improved spectral utilization efficiency with higher order modulation

Modulation	Bits/symbol	E_b/N_o @ BER $= 10^{-3}$
OOK	1	10
QPSK	2	7
16-QAM	4	11
64-QAM	8	17

$$C_{\max} = B. \log_2(1 + SNR) \tag{6.1}$$

The spectral efficiency measured in b/s/Hz is proportional to the SNR in linear units when SNR $\ll$ 0 dB:

$$\frac{C_{\max}}{B} = \frac{\ln(1 + SNR)}{\ln 2} \approx 1.44 SNR \text{ b/s/Hz} \tag{6.2}$$

Conversely, when SNR $\gg$ 0 dB the spectral efficiency becomes proportional to the SNR in dB:

$$\frac{C_{\max}}{B} = \frac{1}{\log_{10} 2}(SNR) = 0.33 SNR_{dB} \text{ b/s/Hz} \tag{6.3}$$

Link adaptation can be supported when more than one modulation scheme is used. Essentially, a high-level modulation is used when the conditions permit. Conversely, low-level modulation is used when the radio link degrades. By combining these as shown in Fig. 6.5, the radio channel is better when a subscriber is close to the basestation (Table 6.4).

Mobility support is based on handover operation, also known as handoff. Handover is an important aspect for mobility because a mobile user has to maintain continuous connection when moving from one cell to another without interruption. The handover can

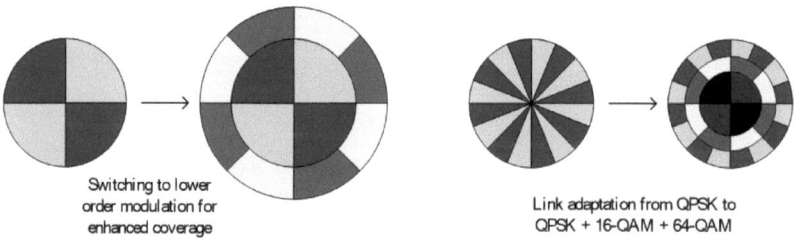

Switching to lower
order modulation for
enhanced coverage

Link adaptation from QPSK to
QPSK + 16-QAM + 64-QAM

Fig. 6.5 Link adaptation

Table 6.4 The first iteration

x_i	μ_1	μ_2	d_1	d_2	Closest cluster centre
(2, 2.5)	(2, 2.5)	(6, 6)	0	5.3151	μ_1
(2, 3)	(2, 2.5)	(6, 6)	0.5000	5.0000	μ_1
(3, 4)	(2, 2.5)	(6, 6)	1.8028	3.6056	μ_1
(6, 6)	(2, 2.5)	(6, 6)	5.3151	0	μ_2
(7, 8)	(2, 2.5)	(6, 6)	7.4330	2.2361	μ_2
(8, 8.5)	(2, 2.5)	(6, 6)	8.4853	3.2016	μ_2

be necessary as a result of mobile subscriber movement, radio link condition changes or cell capacity that makes diversion to an adjacent cell desirable. While handover is a generic feature necessary with any cellular network irrespective of technology, the IEEE 802.16e standard defines two types of handover, namely hard and soft handover. Hard handover works quite simply with the subscriber mobile station (MS) terminates its radio link connection with the first basestation before re-establishing its radio link with the new basestation. It is therefore given the term "break-before-make". In contrast, soft handover works in the form of "make-before-break" such that an MS establishes its radio link with a new basestation before terminating the radio link with the first basestation. The MS can therefore maintain two or more links with multiple basestations, making soft handover inherently faster than hard handover. Soft handover is further divided into two types, Fast BaseStation Switching (FBSS) is a soft handover state where the MS rapidly switching between basestations. The switching operation is fast since the MS switches without carrying out the complete network entry procedure for the new basestation. The other type of soft handover is Macro Diversity HandOver (MDHO), which simply means transmissions are carried out between the MS and multiple basestations.

Summary 6.2 Mobile versus Fixed 6G Networks

1. Networks that support a diverse range of healthcare services often require to provide reliable connection for a combination of both mobile and fixed devices and systems. Network management can differ substantially when support different degree of mobility.
2. Many factors, including modulation scheme and frequency band that are controllable; as well as uncontrollable factors like number of connected devices present and the ability to maintain line-of-sight have substantial impact on operational reliability.
3. Link adaptation can be supported by using multiple modulation schemes within the same system.
4. The same principles and technical considerations concerning optimization between mobility and capacity applies equally for all cellular networks from 4G to 6G.

6.2.3   Telemedicine Networks for Life-Critical Missions

As 6G networks are designed to offer high-speed data transfer with a high degree of mobility through utilization cross-layer optimal resource allocation and combination of spatial multiplexing and beamforming [37], legacy wireless networks that used to deliver only jerky and blurry low resolution video have now evolved into mobile networks that are capable of streaming smooth and clear video at 4 K/60 UHD resolution. Real-time high-resolution video is extremely important in applications like tele-robotic surgery.

Applications ranging from general consumer healthcare monitoring through to emergency medical support can be provided by fast and reliable wireless networks. Given the importance of optimizing system parameters of each and every component of a telemedicine system in order to provide the necessary quality of service (QoS) guarantees required for life-critical applications, a TS has been explicitly developed by the ETSI Special Committee Emergency Communications (EMTEL) to specify a number of objectives and guidelines for emergency notification services [7]. Wireless networks are required to operate with virtually no interruption at all times, irrespective of the harsh operating environment in order to provide reliable connections between the basestation and user nodes. One of the major challenges in providing fast and reliable data transmission is to minimize the probability of error, while maximizing the data throughput under the constraint of very limited available bandwidth. The key to successful system deployment, in a business point of view, is to fully utilize available network resources. In various uncontrollable operating conditions that the system experience, it is vital maintain a certain level of guaranteed quality of service (QoS) in order to support a committed number of subscribers irrespective of changes in the surrounding environment. For example, the effects of persistent heavy rainfall in tropical regions have to be compensated by an appropriate system fade margin through characterization of radio channels under various conditions.

6.2.4   Telemedicine in the 6G Era

The term telemedicine is a broad description of providing medical services with telecommunications. 'Tele' is a prefix for distant, originated from ancient Greek. So, telemedicine literally translates to providing medical services over distance. The first documented case of using telecommunication technology for medicine is probably 'The Radio Doctor' that first appeared in the Radio News magazine in the 1920s [18]. It was made possible by using the Plain Old Telephone Service (POTS) from a simple direct connection between two devices, namely two primitive telephone units. The voice information was subsequently forwarded to the listeners through an AM (amplitude modulation) radio broadcast system. In brief, telemedicine broadly describes the use of telecommunication technology for providing a variety of medical and healthcare services.

Several definitions of telemedicine exist, for example, *wiki* gives its definition as 'Telemedicine is a rapidly developing application of clinical medicine where medical information is transferred through the phone or the Internet and sometimes other networks for the purpose of consulting, and sometimes remote medical procedures or examinations". Other definitions, such as those defined by [7, 8, 12, 13], were also commonly accepted definitions over the past two decades. Regardless of which definition we

Table 6.5 Selected key benefits of telemedicine deployment

Support more types of services	Assistive care, health promotion, emergency medicine
Bring services to more people in more regions	Extend services to rural areas and at sea
Make healthcare reachable by the poor and the elderly	Affordable care with better medical resource utilization
Optimize health for all ages	Ease of use and access
On-scene treatment for medical professionals on the move	Collect real-time patient information and retrieve medical history
Provide preventive care	Continual monitoring, ambient sensing, up-to-date information
Remote rehabilitation monitoring	Reduces hospitalization, improve efficiency, flexible recovery
Chronic disease relief and care	Keep track of medication, provide off-site support
Ascertain service reliability and eliminate human errors	Electronic patient record, automatic prescription and dispensary
Safeguarding patients' information and medical history	Information and network security

take, one key point that they all have in common is that they all describe the use of telecommunication networks for the delivery and support of medical services.

Over the past decade, telemedicine is gaining substantial interests from regulatory authorities, medical professions, caregivers, all the way through to patients. The combined effect of worldwide population growth and aging population gives rise to a significant increase in the demand on medical services. Other factors such as change in lifestyle and environmental pollution are also contributing towards an increase in medical needs [8]. A good telemedicine system would optimize the efficiency of providing a wide range of medical services in terms of both cost and time, save travelling time and make service and support more readily accessible. Some of the benefits of successful telemedicine deployment are summarized in Table 6.5. To address the growing trend of telemedicine deployment in both urban and rural areas throughout the world, faster, more secure and reliable networks that provide enhanced capacity in terms of coverage, data throughput, as well as number of users, is needed for wireless telemedicine development.

6.2.5 Rigorous Requirements

6G networks provide reliable wireless links far exceeds 1 Gbps with no <99.99% availability. The percentage of availability is a measure of system outage time per year. For example, the maximum permissible system outage time is 1–0.9999 within a year to ensure an availability of 99.99%, which translates to approximately 52 min per year when the network is temporarily unavailable. 6G systems provide digital data transmission for many telemedicine applications; here are some of the reasons for the growing interest in these systems:

- Recent development in high speed networking support data rates in excess of 100 Mbps, while many existing users still measure their access speeds in a few Mbps especially with older wireless devices. The main constraint that stops people from getting faster network access is due to bandwidth limitation as its availability is generally too expensive to most consumers. 4G offers a cost-effective access alternative to previous generation networks due to the infrastructure required for deployment is limited to the number of actual subscribers.
- Spectrum is being made available in many countries allocating a significant amount of spectrum outside the very congested 2.4 GHz band; this enables entries of new service providers to deploy cost-effective networks. Also, deregulation and privatization in the telecommunications industry as a global trend in recent years have led to the establishment of many new service providers. 6G goes way beyond 100 GHz that operates in a far less congested radio spectrum.
- Advancements in microwave as well as energy-harvesting circuit designs have led to lower-cost equipment that was previously affordable for pervasive care only in specialized applications. Point-to-multipoint architectures further reduce access costs and system reliability has improved significantly making service more readily available to consumers.
- Mobility is the single most important feature associated with wireless communication networks. Portable devices have been made smaller, lighter, cheaper, and more powerful in recent years due to technical advancements in many areas.
- Scalability is an important feature of wireless systems. Instead of disrupting service for laying expensive cables, it is possible to increase the number of access points by adding more basestations without causing any interruption to existing service. System expansion is generally much easier, faster, and cheaper to perform than wired alternatives.

All these attributes have led to the incentives of optimizing 6G telemedicine system deployments. These access systems can be optimized in terms of utilizing available bandwidth for supporting the maximum number of subscribers, highest data throughput possible, and cover a largest geographical area achievable. The ultimate goal is to minimize operational costs so that subscribers pay less to get data delivered reliably at all times. There are in fact many ways to enhance system optimization due to their open network architecture. While bandwidth utilization is an important factor in effective use of system resources, other parameters such as transmission power, system margin, etc. are also considered when attempting to make a given system as efficient as possible.

The study on optimization of 6G networks for mission-critical telemedicine services opens a new horizon for fine tuning a vast range of operational parameters specific to a given system based on environments and applications. Due to the flexibility of system deployment options, many factors within the system itself as well as external issues must be very carefully considered. In general, the main objective of such system is to deliver high speed digital data over wireless channels to many subscribers. How this can be done and with what resources are used remain important questions to be answered in the system deployment's point of view. The problems that need to be addressed therefore involve adjustment to the system to cope with a number of deployment scenarios as well as considering provisions for operating efficiently irrespective of changes in environment.

Wireless communication systems suffer from a range of degradation factors such as interference and attenuation. These problems arise from both man-made and natural causes and in many cases they are uncontrollable. Optimization of 6G telemedicine systems is found more challenging than most other wireless communication systems because they primarily operate in an outdoor environment subject to a number of atmospheric phenomena. Unlike the indoor environment where factors influencing wave propagation are mostly controllable, many uncontrollable issues that lead to operating uncertainties must be addressed in system deployment. Also, their frequencies of operation in the microwave range make it more susceptible to signal distortion.

Since the obvious goal is to maintain an adequate network QoS standard under given operational constraints such as frequency allocation and available sites for basestation installation. There are many challenges that make system optimization of networks enormously demanding particularly in a medium to long range network backbone. For start, Installation of basestation hubs at optimal sites are not feasible in many cases due to factors such as permission to access, wind loading due to large dish antennas, practicality of performing physical installation (e.g. whether a concrete base is available, water or gas pipes may obstruct installation), etc. It is therefore unlikely that radio hubs can be installed at optimal locations based on calculation. Deficiencies such as cell-to-cell interference and frequency-selective fading can often be controlled internally within the system to minimize any impact caused. However, certain factors such as interference from other systems operating nearby can only be controlled by other parties such as local authorities allocating different frequency bands to different system operators. Further, there are uncontrollable factors such as climate conditions and physical obstacles between the user and the basestation that cannot be controlled at all. For the basestation, the impact of wet antenna may have noticeable effects on the signal [9], covering the antenna may cause undesirable absorption and reflection to the propagating wave while not keeping it dry may affect the measurement results.

6.2.6   Case Study Involving Accident and Emergency Support

Contents Key features of 6G network provide necessary for supporting paramedics attending an accident scene such that information about the injury suffered by a patient can be collected and sent to the hospital in advance. Emergency support personnel at the hospital's A&E department can therefore carry out necessary preparations prior to the patient's arrival thereby providing treatment with minimal time delay. At the same time, on-scene paramedics can also retrieve medical history of the patient through remote access to the patient's electronic patient record (EPR). The telemedicine system block diagram is shown in Fig. 6.6.

This system provides real-time network access to paramedics providing critical care on an accident scene that enables direct two-way communication between an ambulance and the A&E unit of the hospital by an ultra-portable computer linked to central hospital with a 6G network backbone. The hospital database servers store a large amount of information including a patient's medical history. It also facilitates integration of GPS tracking that enables hospital staff to obtain an accurate estimate of the ambulance's arrival time so that necessary preparations can be made with minimal delay. The ability of 6G networks to support high speed vehicle to infrastructure (V2I) communications imply that data can be transmitted from the ambulance during the course of rushing back to the hospital safely while simultaneously transmits various vital signs collected from the patient.

With the importance of this telemedicine system discussed thus far, we investigate the requirements for supporting such system via a 6G network. One of the system's key features is a wearable high-resolution digital camera that captures information about the injury a patient sustains as well as the surrounding environment. The process of delivering the captured video from the on-scene paramedic to the hospital A&E department is shown in Fig. 6.7. In this sub-system, the streaming server processes the incoming raw video data into a format that is suitable for transmission. It is also supports a number of standard playback functions such as fast-forward, rewind, play/pause, etc. Raw video data must be

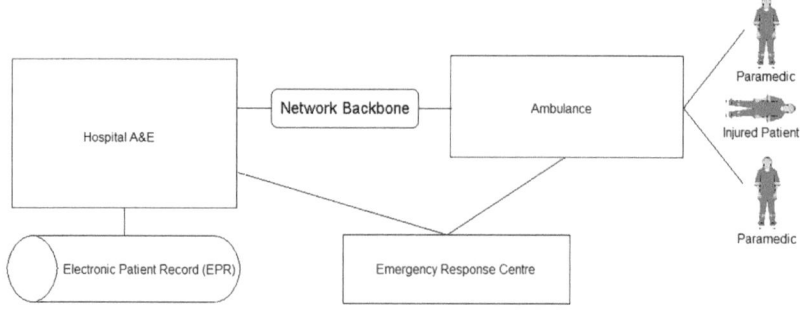

Fig. 6.6 A 6G network backbone supports on-scene emergency medicine

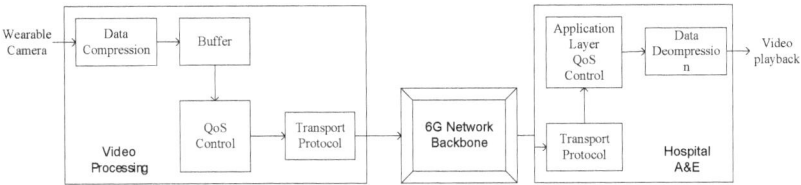

Fig. 6.7 Block diagram of the video transmission sub-system

suitably compressed before transmission to minimize the data traffic volume. Compression is performed based on a tradeoff according to the rate-distortion (R-D) theory [32]. Scalable video compression is often achieved by compressing raw video data into a base layer bitstream, along with some other bitstreams for enhancement layers [17]. Scalability is optimized both for discrete cosine transform (DCT) and wavelet transform with the compressed data stored in the buffer and processed by an application-layer QoS control stage prior to transmission over the network. The client performs the reverse operations of the server to regenerate the video clip.

6G provides a reliable wireless channel for multimedia data distribution. A number of issues associated with the network backbone must be considered in order to make it practical for video streaming applications. Three types of protocols must be addressed here, namely network-layer protocol, transport protocol, and application protocol. The nature of the IP means that application-layer QoS control plays a very important role in ensuring that the quality of video data delivered across the system are maintained at an acceptable level at the receiver.

Procedures involving application-layer QoS control may be viewed from two perspectives: transport and compression [26]. Control techniques viewed from a network transport perspective are employed without reference to any particular video coding scheme (semantically decoupled); whereas Compression perspective is the opposite, where control techniques are applied to video semantics within a compression framework. We focus on techniques applicable to generic video data viewed from a transport perspective. Also, we are interested in sender/receiver-based techniques that make no assumptions on any QoS support that may be offered by network routers. Application-layer QoS control techniques fall into two categories: network congestion control and error control.

Network Congestion Control: Employed to reduce the impact of packet loss and delays caused by network congestion [24]. From a transport perspective, the two major techniques for congestion control are: rate control [34] and rate shaping [26]. While the latter may also be considered from a compression perspective, we focus on techniques to make no reference to specific compression semantics. Rate control reduces network congestion by matching transmission rate with available bandwidth at any given time. In the absence of rate control, packets that are transmitted at a rate higher than what could be supported by the prevailing network bandwidth would be lost. In essence, rate control entails an estimation of the prevailing network condition, which may be performed by the sender and/or receiver. Once the network condition is known, rate shaping (or filtering) is used to constrain the data rate to match the available bandwidth.

Error Control: To reduce the effect of bandwidth limitations, packet loss and delays on the quality of the received video. In the absence of error control, it is virtually impossible to provide acceptable quality of play-out with real-time video data delivered over the Internet. The two major approaches used for error control are retransmission and FEC.

Considered from a transport perspective, retransmission can provide good error rates without incurring much bandwidth overhead because packets are retransmitted only when there are some indications that they are lost. However, retransmission may lead to intolerable delays for real-time video applications. Current solutions to constrain the retransmission delay revolve around adding extended control mechanisms to ensure that retransmitted packets arrive in time. Redundant information is transmitted along with the original information. The lost packets can therefore be recovered directly from the redundant information at the receiver. Unlike retransmission, FEC does not incur additional latency due to retransmission of missing packets. FEC mechanism is more effective when lost packets are dispersed throughout the stream of packets sent from a source to a destination [13]. Therefore, it is well suited to real-time video delivery over the Internet. Considered from a transport perspective, FEC is primarily implemented in the channel coding stage.

An integrated approach that deals with both network congestion control and error control for effective QoS control of streamed video over the Internet is developed based on the above. This integrated approach involves four stages shown in Fig. 6.8 that carries out packet loss analysis, rate control, video stream selection (rate shaping), and FEC.

Its operation is as follows: the receiver collects packet loss information by referring to the packet sequence numbers. For example, if packet n is not received after packets $n + 1$ and $n + 2$ have been received, then packet n is considered lost. Based on this information, the receiver periodically generates and sends packet loss reports to the sender for analysis. For practical purposes, the period is set at once every 20 s. On receipt of every packet loss report, the sender updates the packet loss statistics in regard to the specific receiver that sent the report. This allows for a single sender to deliver video data to multiple receivers at different rates to suit the individual receivers.

In rate control, the network condition experienced by each receiver is estimated and the bandwidth requirement for each receiver is updated to reflect any changes in network

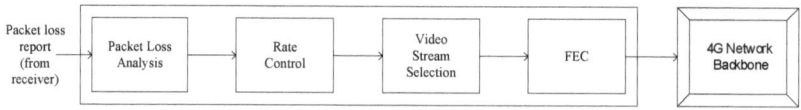

Fig. 6.8 Integrated QoS control

condition since the last adjustment. Using scalable video data compression, the optimal video data stream may be selected for optimal bandwidth usage with respect to each individual receiver. This selection therefore amounts to rate shaping. Finally, channel coding FEC mechanism is performed at the transmitting end, where the number and nature of parity redundant packets are determined according to prevailing network condition.

Summary 6.3 Telemedicine Networks for Life-Critical Missions

1. Network design often entails compromise between performance and reliability, certain system parameters can be adjusted according to variations in operating environment.
2. The requirements for availability and reliability is far more critical for 6G networks support healthcare applications than consumer-oriented applications.
3. The quality-of-service (QoS) of a network can often be optimized through proper network congestion and error control.

6.2.7   Reliability for 6G Telemedicine

As we briefly discussed reliability issues concerning 6G telecare in Sect. 6.3.2, there are a number of important points to note before we proceed further. First, a communication device can fail, possibly due to a bug found in its firmware. Alternatively, there can be a sudden hardware failure. An electronic component malfunctions or a misalignment is found in a mechanical switch. As the operation of mobile communication devices ranging from generic mobile phone to sophisticated medical instruments all increasingly depend on its touch screen display as the user interface (UI), even a simple backlight failure may lead to the entire device being inoperable. There are so many possible causes of failure that we simply cannot cover comprehensively in the text. We also understand that the ever changing operating environment also has a substantial impact on the reliability.

While we do not intend to provide a comprehensive coverage of reliability for wireless telecare, we focus primarily on factors that may influence the reliability of using 6G for telecare support, as perceived by the end users, i.e. healthcare practitioners and patients.

6.2.8   Reliability and Application

Different systems operate under different environments and each system may have its own reliability requirements. To illustrate these differences, we refer to the simple telemedicine system shown in Fig. 6.9, where all entities can be served by a 6G cellular network.

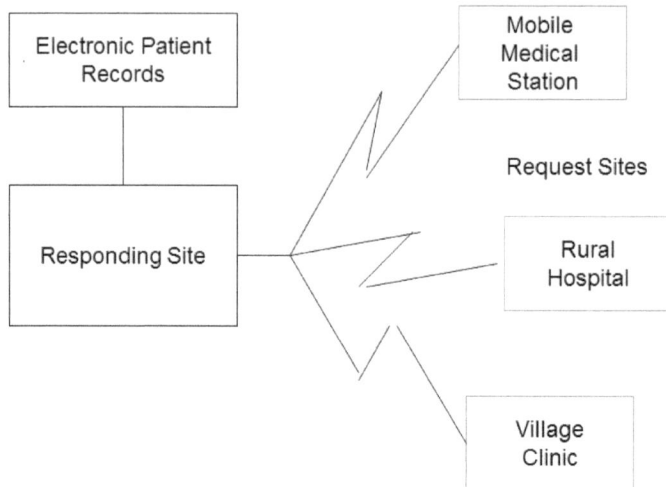

Fig. 6.9 A simple telecare system

6.2.9   Coverage Area Versus Reliability

Wireless telemedicine is becoming increasingly popular due to its ability to readily adjust for new healthcare services with reasonably low initial infrastructure costs by utilizing an existing network backbone. These systems have their merits in the telemedicine industry with vast growth potentials due to a number of reasons:

- Recent extensive development in networking technologies has made mobile and vehicular data delivery faster and cheaper making access cost cheaper and at the same time support more services. 6G is a particularly suitable solution due to its flexible infrastructure required for deployment can be limited to the number of actual subscribers. Point-to-multipoint architectures further minimizes costs and system reliability is consistent with cable networking technologies
- The increasing amount of spectrum being made available along with other networks will enable new data services that in turn will add further to the already growing demand for speed and capacity.
- Deregulation and privatization in the global telecommunication industry has led to the establishment of numerous new service providers. Wireless is the most cost-effective access technology available to these new operators due to its flexibility and relatively low initial infrastructure costs.
- Technological advancements in microwave devices have led to lower-cost equipment that was previously affordable only in specialized applications, eventually making pervasive computing with mobile BWA possible.

Both public and private healthcare providers throughout the world are using wireless to increase efficiency and better manage costs for faster delivery and enhanced patient care. Diagnosis can be carried out using smartphones equipped with various medical applications for fast and secure access to remotely generate laboratory results. The increasing volume of transferred data between healthcare providers and patients requires integrated networks with bandwidth capable of simultaneously supporting different on-demand healthcare applications [35]. Utilizing maximum bandwidth efficiency to provide a versatile system, 6G offers an economical means of setting up a readily scalable network to facilitate growth in demand and emergence of new telehealth services. The flexibility of network deployment based on the basic set provides numerous ways of optimizing system performance. At the transmitting side, incoming signal is modulated and converted into the appropriate carrier frequency and amplified for transmission. The reverse process occurs at the receiving end where the received signal is demodulated and converted back into a signal in the form suitable for distribution to individual end users.

A wireless link is established between the service basestation and each subscriber (i.e. the end-user) offering a means for linking all its subscribers to the network backbone. A basic block diagram showing the system operation is shown in Fig. 6.10.

The basestation of the service node carries out the main task of converting data from the network backbone into signals suitable for wireless transmission and vice versa. It

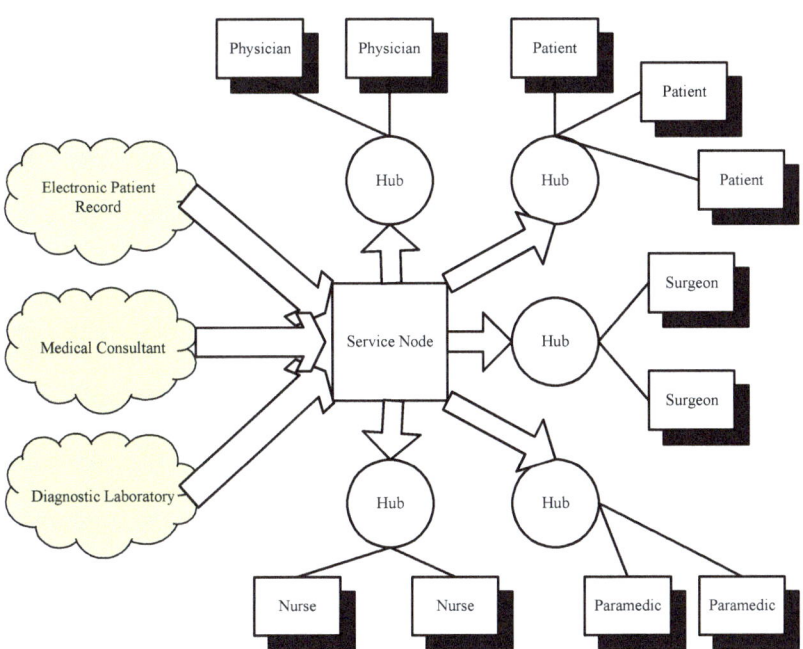

Fig. 6.10 Block diagram of simple 6G telemedicine system deployment

contains the network interface components for link termination and performs function as a wireless modem. The basic building blocks are similar to most alternative technologies such as optical networks and previous generation cellular networks. All these offer a common service point that serves a number of subscribers within a local area.

Earlier study in cell coverage has been shown with an area coverage model approximated by a square [20]. The separation between each hub SH and its coverage area Hi can therefore be approximated as:

$$S_H = \frac{2.r}{\sqrt{2}} \tag{6.4}$$

$$H_i = 2.r^2 \tag{6.5}$$

where r is the range of the unobstructed wireless link. The cell size that ultimately determines the network coverage is primarily dependent on the capacity constraints and placement of end-user antennas. Antennas play a significant part in providing an optimal balance between service reliability and coverage. Cell overlap control is another consideration in antenna deployment [10].

The ubiquitous nature of IP also makes 6G network a suitable solution for achieving scalability in terms of a unified network interface. Because of the high data rate capacity of the radio link between the basestation and the end-user site, packetized voice and video services become much less problematic in terms of their real-time requirements. Therefore, IP can be used as the unified network interface for all the data, voice and video services. Greater scalability is achieved when every service is based on IP instead of several different networking protocols because applications can be added without any network layer protocol changes.

Another important aspect of IP network scalability concerns the peripherals of the overall healthcare network. The increasing demand in such applications as mobile clinics and remote patient monitoring has led to a growth in small remote private networks. These subnets connect seamlessly to the broader telemedicine. So-called extruded subnets and related protocols represent a significant recent development in this area. Extruded subnets connect seamlessly to the Internet via wideband access lines to an organization's main network in a manner that hides the fact that the subnets are remote [6]. Therefore, these subnets are to the outside world very much part of the public network.

Summary 6.4 Reliability for 6G Telemedicine

1. 6G networks will be deployed in a few years' time throughout the world. There will be more types of wearable and implantable connected devices. Supported by a high degree of mobility, travelling on high speed vehicles will not affect the network connection that a user can maintain. All these are made possible due

to a range of enhancements like handover algorithms and hardware design as discussed in the earlier part of this chapter. The fact that 6G offers much higher performance than previous generations of cellular communication standards also enable many telemedicine services made available and accessible.

2. We have looked at the evolutionary path over the past couple of decades from GSM towards the current 5G and upcoming 6G wireless infrastructures such that each iteration is capable of supporting more advanced medical services. Future development from 5G to 6G should see improvements in interconnection and interoperability for supporting a more diverse range of medical applications. In particular, enhancing telemedicine services for mobile health as well as emergency support will certainly allow enhanced medical services to be delivered to more people in more locations.

References

1. M. Aferzon, B. Millman, Excision of rhinophyma with high-frequency electrosurgery. Dermatol. Surg. **28**(8), 735–738 (2002)
2. N. Ansari, P.H. Ho, V. Lau, D.H. Tsang, Special issue on broadband access networks. IEEE Syst. J. **4**(4), 410–412 (2010)
3. Q. Bi, Ten trends in the cellular industry and an outlook on 6G. IEEE Commun. Mag. **57**(12), 31–36 (2019)
4. R. Bruno, M. Conti, E. Gregori, Mesh networks: commodity multihop ad hoc networks. IEEE Commun. Mag. **43**(3), 123–131 (2005)
5. L. Caviglione, A simple neural framework for bandwidth reservation of VoIP communications in cost-effective devices. IEEE Trans. Consum. Electron. **56**(3), 1252–1257 (2010)
6. C. de Launois, A. Bonnet, M. Lobelle, Connection of extruded subnets: a solution based on RSIP. IEEE Commun. Mag. **40**(9), 116–121 (2002)
7. ETSI TS 102 182 V1.2.1, Emergency Communications (EMTEL); Requirements for communications from authorities/organizations to individuals, groups or the general public during emergencies (2006)
8. B. Fong, A.C.M. Fong, C.K. Li, Telemedicine technologies: information technologies in medicine and telehealth. John Wiley & Sons (2011)
9. B. Fong, P.B. Rapajic, G.Y. Hong, A.C.M. Fong, Factors causing uncertainties in outdoor wireless wearable communications. IEEE Pervasive Comput. **2**(2), 16–19 (2003)
10. B. Fong, N. Ansari, A.C.M. Fong, G.Y. Hong, On the scalability of fixed broadband wireless access network deployment. IEEE Commun. Mag. **42**(9), S12–S18 (2004)
11. B. Fong, K.F. Tsang, C.K. Li, Personalized elderly assistive home care using 3G networks. J. Adv. Inform. Technol. **4**(3), 136–141 (2013)
12. H.Y. Gu, C.Y. Yang, B. Fong, Low-complexity centralized joint power and admission control in cognitive radio networks. IEEE Commun. Lett. **13**(6), 420–422 (2009)
13. S.M. Huynh, D. Parry, A.C.M. Fong, J. Tang, Novel RFID and ontology based home localization system for misplaced objects. IEEE Trans. Consum. Electron. **60**(3), 402–410 (2014)

14. F. Izzo et al., Radiofrequency ablation and microwave ablation in liver tumors: an update. Oncologist **24**(10), e990–e1005 (2019)

15. J. Karamachoski, L. Gavrilovska, Framework for next generation of digital healthcare systems, in *International Conference on Future Access Enablers of Ubiquitous and Intelligent Infrastructures* (Springer, 2019), pp. 12–24

16. V. Mancuso, S. Alouf, Reducing costs and pollution in cellular networks. IEEE Commun. Mag. **49**(8), 63–71 (2011)

17. Y. Mei, L. Li, Z. Li, F. Li, Learning-based scalable image compression with latent-feature reuse and prediction. IEEE Trans. Multimedia **24**(9), 4143–4157 (2021)

18. G.T. Moore et al., Comparison of television and telephone for remote medical consultation. N. Engl. J. Med. **292**(14), 729–732 (1975)

19. T. Nakamura et al., Trends in small cell enhancements in LTE advanced. IEEE Commun. Mag. **51**(2), 98–105 (2013)

20. P.B. Papazian, G.A. Hufford, R.J. Achatz, R. Hoffman, Study of the local multipoint distribution service radio channel. IEEE Trans. Broadcast. **43**(2), 175–184 (1997)

21. S. Parkvall, et al., LTE-advanced-evolving LTE towards IMT-advanced, in *2008 IEEE 68th Vehicular Technology Conference*, September (IEEE, 2008)

22. R.V. Prasad, P. Pawelczak, J.A. Hoffmeyer, H.S. Berger, Cognitive functionality in next generation wireless networks: standardization efforts. IEEE Commun. Mag. **46**(4), 72–78 (2008)

23. R.G. Saez, N.M. Marques, RF power amplifier linearization in professional mobile radio communications using artificial neural networks. IEEE Trans. Industr. Electron. **66**(4), 3060–3070 (2018)

24. Y. Tian, K. Xu, N. Ansari, TCP in wireless environments: problems and solutions. IEEE Commun. Mag. **43**(3), S27–S32 (2005)

25. A.A. Toor et al., A robust systematic approach for ensuring optimal telecom service delivery. IEEE Commun. Mag. **58**(8), 49–53 (2020)

26. D. Wu, Y.T. Hou, Y.Q. Zhang, Transporting real-time video over the Internet: challenges and approaches. Proc. IEEE **88**(12), 1855–1877 (2000)

27. T.C. Wu et al., Prehospital utility of rapid stroke evaluation using in-ambulance telemedicine: a pilot feasibility study. Stroke **45**(8), 2342–2347 (2014)

28. Y. Yang, H. Hu, J. Xu, G. Mao, Relay technologies for WiMax and LTE-advanced mobile systems. IEEE Commun. Mag. **47**(10), 100–105 (2009)

29. A. Yeung, C. Wu, AI for elderly dietary care: a cyber-physical system for nutrition management. IEEE Smart Cities Newsl. **3**(8) (2021)

30. F. Yilmaz, On the relationships between average channel capacity, average bit error rate, outage probability, and outage capacity over additive white Gaussian noise channels. IEEE Trans. Commun. **68**(5), 2763–2776 (2020)

31. H. Yokota, A. Idoue, Emerging standards for mobility management in next-generation All-IP networks, in *Invited Paper, Third International Conference on Mobile Computing and Ubiquitous Networking (ICMU 2006)*

32. H. Yuan, et al., Hybrid distortion-based rate-distortion optimization and rate control for H. 265/HEVC. IEEE Trans. Consum. Electron. **67**(2), 97–106 (2021)

33. J. Zhang, N. Ansari, On OFDMA resource allocation and wavelength assignment in OFDMA-based WDM radio-over-fiber picocellular networks. IEEE J. Sel. Areas Commun. **29**(6), 1273–1283 (2011)

34. T. Zhang, S. Mao, Machine learning for end-to-end congestion control. IEEE Commun. Mag. **58**(6), 52–57 (2020)

35. Y. Zhang, N. Ansari, H. Tsunoda, Wireless telemedicine services over integrated IEEE 802.11/WLAN and IEEE 802.16/WiMAX networks. IEEE Wirel. Commun. **17**(1), 30–36 (2010)

36. Y. Zhou, L. Tian, L. Liu, Y. Qi, Fog computing enabled future mobile communication networks: a convergence of communication and computing. IEEE Commun. Mag. **57**(5), 20–27 (2019)
37. T. Zugno, M. Drago, M. Giordani, M. Polese, M. Zorzi, Toward standardization of millimeter-wave vehicle-to-vehicle networks: open challenges and performance evaluation. IEEE Commun. Mag. **58**(9), 79–85 (2020)

Use Cases and Reliability Optimization for 6G Healthcare Systems in Smart Cities Integration

Cellular communication networks serve as a vital infrastructure backbone that links different areas such as smart health, education, energy, transportation, environmental and resource management. The development of 6G systems would link various connected devices and monitoring systems for effective disease-spread simulation analysis that enables health resource management in tackling infectious disease outbreaks; monitor pollutions and to support smart transportation management for efficient route planning to ease congestion and minimize toxic fume emission to provide a healthier and safer environment across a smart city.

Central to the successful implementation of a smart city is a reliable, scalable, and is economical to maintain [34]. Communication networks provide the fundamental function within a smart city infrastructure that links many different types of systems together. Assessing and assuring maximum reliability of such communication networks require thorough analysis of various performance metrics under different operating environments.

7.1 Key Performance Indicators (KPIs) in Connection Reliability

The operational cost for network access to be provided to a certain number of users is primarily determined by parameters such as data throughput, transmission power efficiency as well as coverage in terms of both total service area and number of devices per base station. The data required throughput for a given device can be computed based on statistical analysis of past usage per device from the number/type of services a given device supports, e.g. health monitoring and hazard detection.

© The Author(s), under exclusive license to Springer Nature Switzerland AG 2025 121
V. Sai et al., *6G Enabled Healthcare Systems*, Synthesis Lectures on Engineering, Science, and Technology, https://doi.org/10.1007/978-3-031-73849-4_7

Mobile communication devices are particularly prone to wave propagation issues such as Doppler effects and blockage resulting from different kinds of physical obstacles the exhibit varying reflection and absorption properties [30].

Operational reliability relies on maintaining a radio link under various operating environments. Various physical obstacles can cause reflection, diffraction, and shadow fading to propagating radio waves. Signal attenuation, the weakening of signal strength over distance travelled, due to rain droplets can be very significant that is particularly problematic at radio frequencies used by 6G carriers.

Wearable and implantable medical as well as consumer healthcare devices are becoming increasingly popular with emergence of new technologies that support more comprehensive healthcare services, making them suitable for accessing medical services when the patients move around. 6G networks offer many advantages over 5G or other earlier generations of cellular communication systems in areas such as scalability, lower latency and delay; as well as higher degree of mobility within a locality. All connected devices worn by a patient need to remain connected when the patient moves around. The wireless link has to be very reliable in order to provide a sustainable connection between the transmitter and the receiver irrespective of the ever-changing operating environment. Various uncontrollable parameters like rain-induced attenuation and signal degradation due to physical objects have to be dealt with by dynamically adjusting parameters such as power and gain.

Implanted medical devices, in particular, that operate with very low power consumption have implementation constraints limited by its physical size and its ability to receive the wireless signals under different environments that needs to penetrate through the patient's body. While multipath is much less of an issue with 6G when the carrier frequency is high enough, effects of rain-induced attenuation and depolarization become more severe as the operating frequency increases.

7.1.1   Cell-To-Cell Interference

Since the concept of microcell was implemented in the 5G era, interference between adjacent cells becomes a more serious issue [20]. The severity of interference induced by nearby cells can cause substantial degradation to the signal within a given cell. The actual operating environment needs to be studied in order to reduce the impact of cell-to-cell interference to a given cell. Such study entails identification of any potential sources of interference across the cell and also to analyze the potential effects of interference from adjacent cells.

It is often necessary to take measurements at arbitrary locations within the cell so to compare the relative signal strength from the particular base station that serves the cell of concern verses those that are originated from other nearby cell. This is particularly problematic when a direct line-of-sight (LOS) path exists between the base station of another cell and the receiver that takes measurement of the signal strength [18].

7.1.2   Basestation Placement

Geographical issues often restrict the practicality of installing base station at the best location for minimal cell-to-cell interference. There are situations such as the location where optimal location is found may not be a suitable site or no legal permission is granted for installing the base station at that particular location. Most base stations are therefore not installed in their optimal positions thereby leading to imprecise patterns of cells [50]. Certain base station installation locations can exhibit a higher degree of co-channel interference than another location within the same cell [37].

Figure 7.1 demonstrates an example of which a cell is being split into three sections. In this illustration, a base station uses 8 hubs to serve 5 sites grouped into 3 sectors. *Sector A* contains the highest concentration of subscribers and this illustration has 3 subscribers that is simultaneously served by the same basestation. Each of the 8 hubs should use a slightly different carrier frequency or use a different polarization under a scheme that utilizes 4 pairs of frequency sub-channels in order to eliminate any interference between each other. Another observation is that while each hub is, in theory, optimally placed with direct LOS maintained for all subscribers, movement of subscriber nodes can impact signal reception to varying degrees. Moreover, certain limitations make the optimal placement of hubs impractical, these could be due to physical or licensing issues. At least 2 hubs should be allocated to serve each sector.

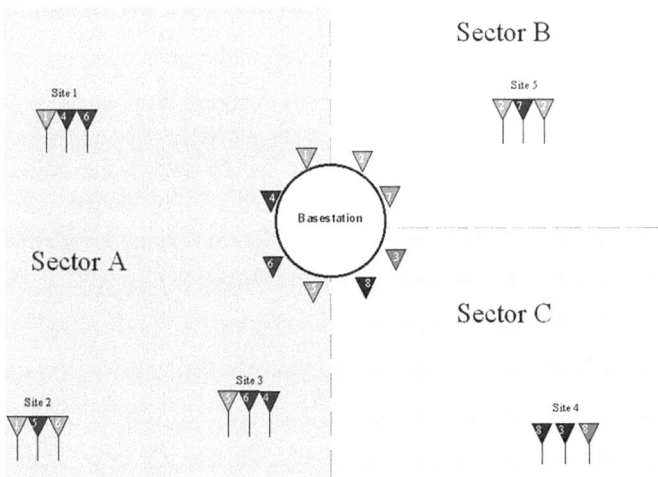

Fig. 7.1   A base station serving three sectors with eight hubs

7.1.3 Weather Phenomena

Uncontrollable weather conditions such as rain, snow and absorption caused by water vapour can cause problems to propagating waves due to attenuation and depolarization. The extent of signal degradation is primarily dependent on the carrier frequency and rainfall rate at the time, this is particularly problematic when 6G goes way above carrier frequency of current 5G signals. Earlier study of long-term statistics has suggested that rain-induced attenuation is the most influential degradation factor on the availability and reliability of radio links that operate above 10 GHz [12].

Rain-induced attenuation could substantially reduce the signal strength even a short path at carrier frequencies used in 6G, depolarization is especially problematic when a system utilizes alternately polarized signal paths where orthogonally polarized signals are often deployed to enhance bandwidth efficiency [19]. Rain-induced depolarization causes cross-polarization interference that reduces the polarization separation between the two signal paths of orthogonal polarizations when horizontally polarized signal suffers a higher degree of depolarization [13].

One of the most challenging attributes in supporting healthcare applications using 6G networks is to maximize quality of service (QoS) as wireless networks that operate in an outdoor environment is often under harsh environments, including weather phenomena and various physical issues like reflection, diffraction and scattering [27].

To illustrate the extent of severity caused by rain, we take a closer look at a comparison between carrier frequencies at 10 and 26 GHz, where rain attenuation is analyzed as a function of data rate. Such performance indicator can provide insights into the compensation necessary for combat the effects of rain attenuation so that fade margin requirements can be established for optimal system design and implementation.

The operational performance of a point-to-point wireless connection is analyzed between a transmitter–receiver pair to evaluate the effects of rainfall on the link availability by comparing the selected frequencies when transmitting raw data at different bit rates, i.e. only performance at the physical layer is analyzed from the Open Systems Interconnection (OSI) model's perspective such that comparison is made between transmitting at 20 and 40 Mbps using carrier frequencies of 10 and 26 GHz. The maximum link coverage range is measured for a bit error rate (BER) of 10^{-6}, which indicates the likelihood of receiving one corrupted bit within a block of 1 million bits being transferred statistically, with a system gain G of 110 dB. Where G is a calculated as a ratio of the transmitted power to the receiver sensitivity given by:

$$G = \frac{k}{R_B} \tag{7.1}$$

where k is an arbitrary constant and R_B is the data transmission rate measured in number of bits per second (bps). Link coverage is primarily determined by factors such as both transmitter and receiver's antenna gains measured in dBi and the rate of rainfall in mm/hr.

The reduction in link coverage range is illustrated as a function of the rate of rainfall. Figure 7.2 indicates that 10 GHz provides a longer range than 26 GHz under identical operating conditions and the difference between the two selected carrier frequencies reduces when the rain becomes heavier. It is also noted that the link coverage is slightly reduced when the data rate is doubled but not by a substantial amount. The range ratio d' is computed as:

$$d' = \sqrt{\frac{R_{\max}}{R_{\min}}} \tag{7.2}$$

Fig. 7.2 Link coverage in km as a function of rate of rainfall in mm/hr, solid line indicates the performance at a data transmission of rate 20 Mbps whereas the dotted line is at 40 Mbps

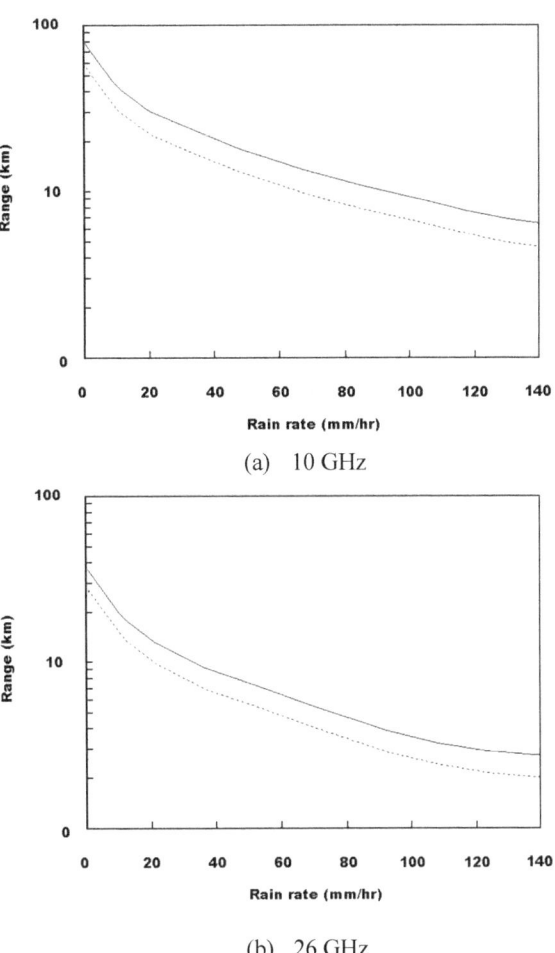

(a) 10 GHz

(b) 26 GHz

where R_{max} and R_{min} denote the maximum and minimum data rates.

To analyze the impact of rainfall on cross-polarization effects, the dual cross-coupling between two signals of orthogonal polarization is measured to compute the cross polarization diversity (XPD). XPD is typically used to quantitatively measure amount of signal degradation, which is determined by the degree of coupling between orthogonally polarized signal paths [18]. XPD results in a 10% reduction in coverage due to cell-to-cell interference such that:

$$XPD = -20\log\big(l.\cos^2 \varepsilon.k.\exp(-2\sigma^2)\big) \qquad (7.3)$$

where

$$k = \sqrt{\alpha^2 + \theta^2} \qquad (7.4)$$

l measures the path length in km, ε is the elevation angle between the transmitting base station and the receiving node, σ is the standard deviation of the canting angle distribution, α is the attenuation in dB and θ is the phase shift in degrees. Figures 7.3 and 7.4 plot the phase difference between the two orthogonally polarized signals and XPD, respectively. The effects of heavy rainfall reduce the link availability due to cross-polarization interference that subsequently diminishes the polarization separation between the two signal paths.

Summary 7.1 Key Performance Indicators (KPIs) in Connection Reliability

1. Connection reliability is about optimizing a collection of network parameters, which often result in a compromise on certain attributes.
2. Physical obstacles along the wave propagating path can cause reflection, diffraction, and shadow fading to propagating radio waves.
3. Rain droplets can cause substantial signal attenuation and depolarization to signals especially with carrier frequencies used by 6G systems.

7.2   Operation of 6G Networks in a Smart City Environment

The cellular network infrastructure of a smart city provides the fundamental underlying link to a variety of services and systems. The important objective of ensuring reliable operation of the 6G network in a smart city is to maintain an adequate level of QoS so that network resources are sufficient for supporting different services. Design and implementation of 6G networks in a smart city environment entails integration of different systems that depending on each other. For example, health management for patients with chronic obstructive pulmonary disease (COPD) would need simultaneous support from air quality surveillance system. Further extending this example of managing COPD patients'

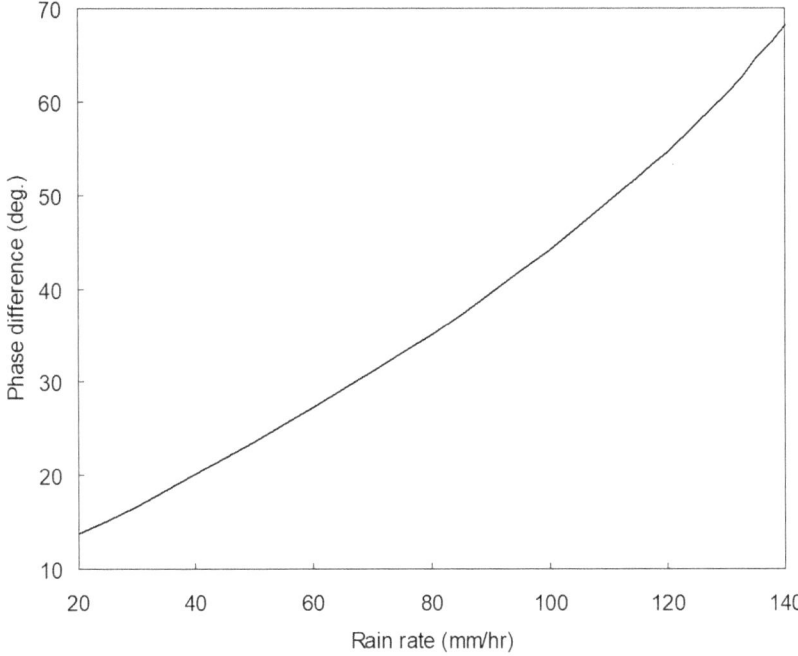

Fig. 7.3 Relative phase of the two orthogonally polarized signals as a function of the rate of rainfall

health from the smart city context into smart home integration, various IoT-enabled consumer electronics devices will work together to provide the best possible environment to minimize the health issues concerning COPD for the patient as shown in Fig. 7.5.

Certain system parameters can be adjusted to improve overall system performance, for example, coding and data compression schemes can enhance the efficiency of data transmission. Efficient implementation of mechanisms such as error correction and equalization techniques to minimize inter-symbol interference (ISI) as well as utilization of space and time diversity are all measures that can improve operational reliability [26].

7.2.1   Pervasive Health Support

The discussion on supporting pervasive healthcare can commence by taking a look at the common example of using a multi-purpose smartwatch. In its most primitive form, a health monitoring system can be made up of a connected smartwatch that measures various health signs of a patient, such as blood glucose level, electrocardiogram (ECG), heart and respiratory rates. More advanced systems can support paramedics that attends to a patient to collect information such as extent of injury and to obtain information

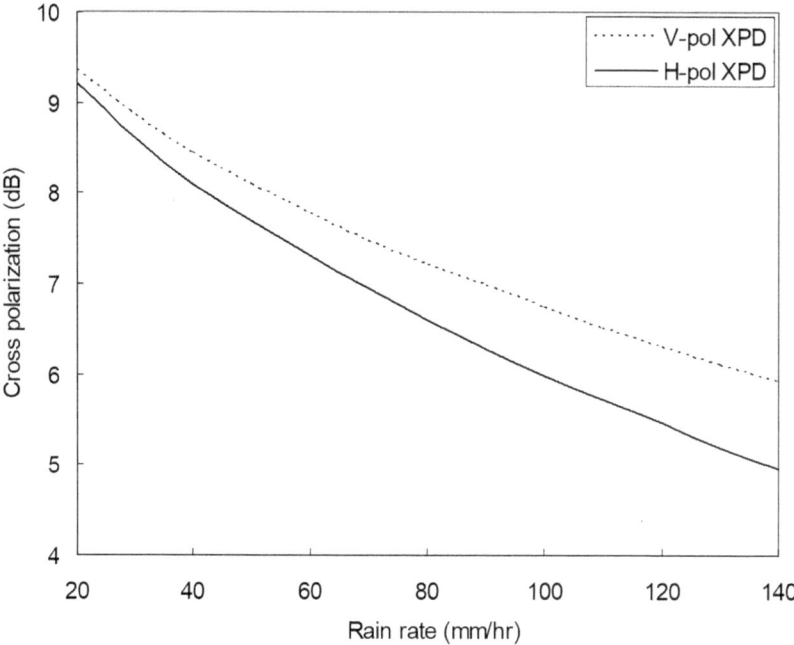

Fig. 7.4 Cross polarization diversity (XPD) per km of path length

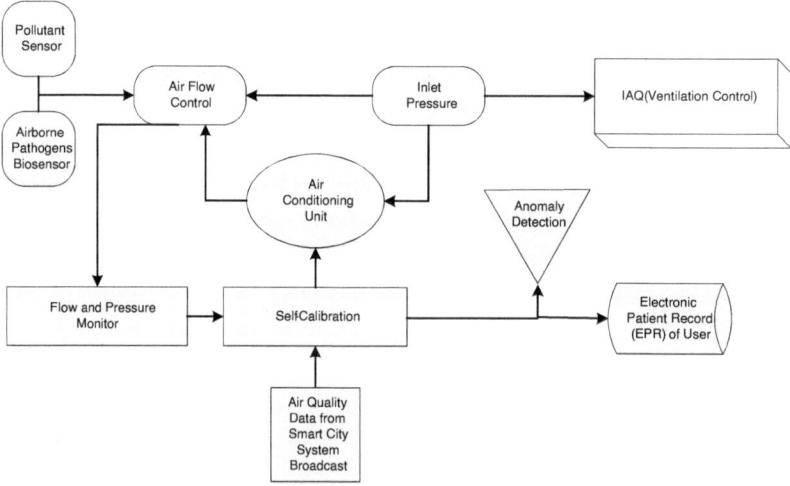

Fig. 7.5 6G network backbone that provides important connection from smart city and smart home integration in a case study for chronic obstructive pulmonary disease (COPD) health management

about the patient's medical records, all within a smart ambulance system [18]. Combining pervasive connectivity and AI technologies such as self-cognizant prognostics, 6G's ultra-reliable low-latency communication (URLLC) and massive device connectivity can support a wide range of healthcare and medical applications in smart cities. The ability to support reliable communication between devices for real-time data sharing and decision-making can enhance energy management and to make smart ambulance more efficient. Additionally, the enhanced bandwidth and data throughput can support advanced applications such as augmented reality (AR) and virtual reality (VR), making medical training more immersive and interactive in areas such as emergency medicine and remote surgery. This holistic connectivity will not only enhance service reliability and efficiency but also create a more integrated and responsive smart healthcare environment.

7.2.2   Environmental Health Support

Behind the economic prosperity resulting from industrialization over the past two centuries is opportunity cost such as environmental damage and related health issues [4]. First, industrialization causes air pollution and discharge of toxic substances that induces a range of nasal, bronchial and sinus diseases [1]. Another important contributing factor to worsening environmental pollution in urban cities due to rapid industrialization that drives many people from rural areas into urban areas that put even more pressure to the healthcare system of metropolitan cities [53].

As industrial revolution changed the landscape of manufacturing and mining in America around the dawn of the nineteenth century, fossil fuel burning and toxic gas discharge have significantly accelerated that in turn create health related issues such as air pollution and acid rain. Although there is no doubt that industrialization directly causes negative environmental impacts on people's health, the trend of industrialization spreads eastwards into Asia from the early time of post-war era. For example, the manufacturing of lead-acid batteries saw its shift gradually from the western world to the far east that also brings along the associated health hazards (Van der Kuijp). A variety of chronic diseases were prominent throughout the industrial revolution as it changed the way of how people worked, and so did the way of disease spread [23]. Environmental related diseases such as cholera, typhus and tuberculosis that roamed across industrialized nations during the mid-nineteenth century are no longer of significant danger due to cleaner water and enhanced preventive healthcare, the threat of disease related to industrialization is far from over [22].

While environmental pollution affects virtually everyone in a city, addressing any potential health hazards caused either directly or indirectly by pollution require efficient and reliable healthcare services. Funding is one major issue in many national health care systems in addition to shortage of nursing staff, technologies that can enhance efficiency and reduce the risk of human errors by using various artificial intelligence (AI) healthcare solutions [2]. Providing accessible and affordable healthcare services across a smart city relies on a reliable communication network that spans across supply chain management and healthcare providers [52]. The capacity and performance of 6G networks provide an important means of connecting various entities in healthcare and environmental sensing.

Looking at a classical example, this kind of vast poisoning may sound familiar in America with a classic example where fluoride-related toxic industrial pollutant has been discharged continuously for half a century, a vast amount of fluoride waste is discharged into the water ecological system [11]. According to Environmental Protection Agency's (EPA) estimation, as much as half a million tons has been released into rivers and oceans by industrial plants in the US a year,with a further 150,000 tons into the atmosphere. Sources of fluoride contamination include metal smelters, petrochemicals, agricultural fertilizers, glass and ceramics manufacturing plants. Fluorides' negative impacts on human and animal lives had been known as early as in the 1930s [45]. This widespread problem also affects other industrialized nations, such as the Meuse Valley of Belgium incident in 1930 [33] where thousands were reported to be poisoned with 60 fatalities, air pollution due to fluoride was blamed by the authority. Many lawsuits were filed after the World War II throughout America for health damages related to fluorite pollution.

In particular, highly toxic dioxins such as chlorinated dibenzo-p-dioxins (CDDs), chlorinated dibenzofurans (CDFs) and certain polychlorinated biphenyls (PCBs) are known to be produced by diesel trucks' emission systems. Further, nitrogen oxides are discharged in vast quantities causing a reddish-brown layer of pollutant over many metropolitan cities. Toxic chemical discharge is also an increasing environmental issue associated with civil activities such as manufacturing and exhaust from combustion engines. Heavy metal is accumulated in the liver catalyzed by alcohol, putting those who drink alcohol on a regular basis on higher risks of liver failure. Dealing with these health issues would require collective efforts in the fronts of health and environmental monitoring, which heavily depend on numerous connected sensors both worn by patients and installed across the smart city infrastructure.

7.2.3   Preventive Care

6G networks are designed to connect a vast number of devices with speed and reliability. With a smartphone serving as a console coupled with a diverse range of other devices and sensors, a user's health and safety can be monitored in relation to the surrounding environment and potential hazards. To illustrate what can be accomplished beyond what current multipurpose smartwatches can do, we take a look at a case study where the system can monitor alcohol consumption both for the prevention of drink-driving and disease control. While the former can easily disable the ignition of a motor vehicle [6], the latter would entail more connected devices and sensors working together.

Excessive consumption of alcohol contributes to liver disease including alcoholic liver disease (ALD) and chronic hepatitis C virus (HCV) infection. Since alcohol-related cirrhosis develop gradually, prognostic analysis becomes important prior to symptoms such as tender liver enlargement, ascites or jaundice are detected [38]. The combined effect of hepatocellular carcinoma (HCC) and hepatitis is set to significantly increase the risk of chronic liver disease. HCC is usually detected by routine screening or at a later stage when symptoms such as upper quadrant pain and sudden weight loss so that body weight monitoring can detect any sign of anomaly [32]. In addition to alcohol's impacts on liver damage, heavy drinking also increases the risks of breast cancer [49], where a combination of mammography screening and public awareness promotion can provide solutions for impeding early treatment.

Summary 7.2 Operation of 6G Networks in a Smart City Environment

1. 6G networks serve as the fundamental communication backbone within a smart city that provide vital linkage between different service sectors.
2. Integrating with sensing networks provide support for a wide range of pervasive and preventive healthcare services across smart cities.

7.3 6G Telemedicine Applications

Cellular networks that utilize high carrier frequencies operate under harsh environmental conditions. As discussed earlier in Sect. 6.4, factors like weather condition and vehicle travelling at high speeds can make signal reception difficult. For example, depolarization caused by rain droplets can have a substantial impact the propagating signal. Any scheme that addresses issues in signal distortion would require minimal computational complexity since many healthcare applications are supported by very small wearable and implantable devices so that power consumption has to be kept to a minimal.

7.3.1 Equalization

Equalization in wireless communications addresses the problems associated with inter-symbol interference (ISI) prior to symbol detection [24]. The equalizer runs its signal processing algorithms so that demodulation can be properly performed to recover the original data from the received signal (Fig. 7.6).

At carrier frequencies used in 6G systems which is more suitable for regions where heavy rainfall is not normally persistent given that the shorter path length of propagating waves at these frequencies are less prone to multipath mitigation particularly when used in conjunction with narrow beamwidth antennas by the receivers [15].

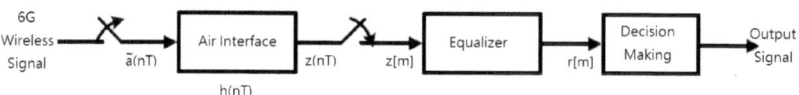

Fig. 7.6 Basic operation of an equalizer

Equalization is particularly critical for receivers in 6G systems due to the greater depolarization effects caused by rain droplets. An equalizer can reduce the impact of ISI using spatial and temporal information by analyzing the received signal [14]. Owning to the low-power requirements for implantable devices in particular, the common approach of deploying adaptive array with temporal filter is not desirable for telemedicine applications hence smart antennas can also be deployed [8].

Smart antenna systems are also deployed for combatting the impact of interference in cases where channel propagation characteristics properties are known [28]. This entails modelling of spatial signatures for selective approaches in both the transmitter and receiver ends, which is generally time-invariant [31], such that the direction of arrival (DOA) of a given signal path can be determined by the spatial signatures.

Analyzing the received signal in both space and time can provide important insights into enhancing signal recovery such that the impact of interference during transmission can be reduced. Spatial equalization is performed once the signal arrives at the received. The wave field e, expressed as a function of temporal position t and spatial location r of the received signal is computed from:

$$\left(\nabla^2 - \frac{1}{c^2}\frac{\partial^2}{\partial t^2}\right)e(t, r) = 0 \tag{7.5}$$

where c is the propagation velocity of the carrier wave and ∇ is given by:

$$\nabla\Phi = a_x\frac{\partial\Phi}{\partial x} + a_y\frac{\partial\Phi}{\partial y} + a_z\frac{\partial\Phi}{\partial z} \tag{7.6}$$

The wave field can then be computed from the linear space–time model:

$$y = \iint h(t, \tau, r, r')x(\tau, r')\frac{dr'}{dr} \tag{7.7}$$

where h denotes the channel's temporal and spatial wave field response at position r subject to delay τ at spatial location r'. The main advantage of such spatial equalization implementation is that being independent of the maximum time delay experienced by the received signal would translate to a reasonably simple process thereby reduces the power requirement. Sensitivity to DOA would be best dealt with by using temporal equalization being independent to the difference in DOA statistics [54].

The frequency response of the spatial–temporal equalizer is plotted in Fig. 7.7, such that the direct LOS path is set as a reference point at 0-degree angle of arrival (AOA). The signal is generated by the following impulse response:

$$f(k) = \frac{1}{N} \sum_{l=1}^{N} |f_l(k)| \tag{7.8}$$

The bit error rate (BER) performance with five taps as a function of the input signal quality is evaluated in Fig. 7.8 which shows that its performance enhancement saturates at near a BER of 7×10^{-5}.

Enhancing efficiency and reliability is one of the main objectives of smart cities. While 6G systems have the capacity of supporting an entire fleet of smart ambulances across a metropolitan area, the actual implementation entails thorough study of certain network parameters to ensure the necessary system and link margins can be maintained.

Fig. 7.7 Frequency response of the equalizer

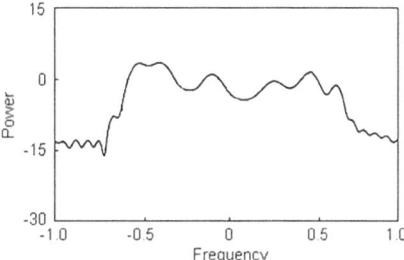

Fig. 7.8 Bit error rate (BER) performance

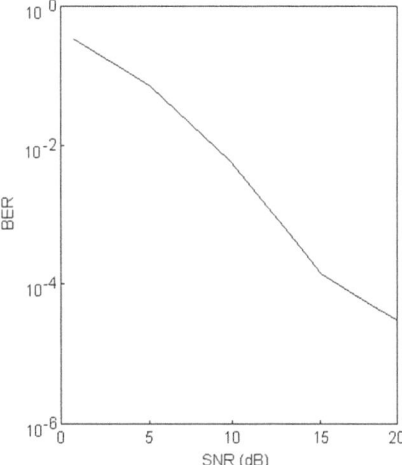

7.3.2   Wave Propagation in Urban Areas

Cellular communications are particularly prone to path loss that can be affected by parameters such as rain-induced attenuation as discussed earlier, antenna position, diffraction loss due to stationary and moving physical obstacles (e.g. buildings, trees, vehicles, etc.). Path loss can be very significant in 6G systems with carrier frequencies in excess of 100 GHz [42]. The link margin has to be sufficient to combat the effects of path loss. The link itself has to provide continuous access to each of the ambulance. In the context of a smart ambulance, it has one set of network access equipment to support paramedics operating inside or near the ambulance, and another external link to the hospital.

Since the 5G era, microcell has been the implementation goal such that cells are more closely packed with smaller coverage per cell. This in turn reduces radio hub spacing. Moving from 5G into 6G has a substantial implication on wave propagation due to much shorter carriers' wavelength, LOS becomes a more important factor to be considered in 6G deployment. Such LOS constraint can impact both the scalability in terms of its geographical coverage and signal interference [46]. System implementation would require the use of physically small and high gain directional antennas to be mounted on top of the smart ambulance in order to reject multipath signal components that arrive from directions other than the direct LOS path.

Radio hub coverage can be approximated with an area of square shape optimally placed at hub separation S_H calculated from:

$$S_H = \frac{2.d}{\sqrt{2}} \tag{7.9}$$

Summary 7.3 6G Telemedicine Applications

1. Equalization schemes can be deployed to optimize operational reliability under different network conditions. Equalization is particularly critical for receivers in 6G systems due to the greater depolarization effects caused by rain droplets with carrier frequencies in excess of 100 GHz.
2. Smart antenna systems are also deployed for combatting the impact of interference in cases where channel propagation characteristics properties are known.
3. The link margin has to be sufficient to combat the effects of path loss.
4. Line-of-sight requirements can impact both the scalability in terms of its geographical coverage and signal interference

7.4   Hardware Reliability Assessment and Optimization

Making communication devices, circuits and systems operate reliability under different use conditions is extremely important in the overall reliability of any 6G cellular networks. This section wraps up our discussion by taking a look at considerations in hardware devices reliability.

7.4.1   Failure Mode Analysis

Cellular networks consist of network equipment across the entire smart city infrastructure that support various end users' consumer electronics and computing devices. Any network equipment failure would impact the target network availability. Many base stations across a smart city are mounted in exposed locations as these are often open areas that provides better coverage. This could have a negative impact on accelerated aging of certain electronic components [25]. Electronic component life cycle loads can be significantly affected by operating environmental conditions such as temperature, humidity, vibration, shock, utilization duration and frequency [47]. Sampling network performance data is a necessary procedure in order to determine the condition of individual base stations installed in different locations of the smart city. The daily duty cycle of the components needs to be studied, the actual working conditions can vary substantially in different locations as the harshness of the environment can have different implications on component health degradation. For example, under extreme conditions such as under hot summer weather and heavy downpour, may demand further field tests in order to establish the actual environmental condition of the systems under the worst servicing condition. Wet antennas in exposed locations can also be a factor that affect network reliability [39].

7.4.2   Laboratory Versus Field Tests

Cellular
In a laboratory environment, it is often desirable to conduct accelerated life tests, such that sample PCBs from a device or system will be placed in a controlled environment with high temperature as well as vibration shocks (Muhammad, 2020). Components such as IC, capacitor and relay are generally found to exhibit the highest failure rate among network equipment. A survey of components that fail across a cellular network across the Greater Auckland area across a three-year period between 2017–19 is plotted in Fig. 7.9 where X-axis shows the abbreviation of each component type. ICs contributed to some half of all component failures, copper traces and relays were also notable failures.

The cross-section images of two sample PCBs from different base stations are shown in Fig. 7.10. The copper thickness can deplete over time particularly when mounted in

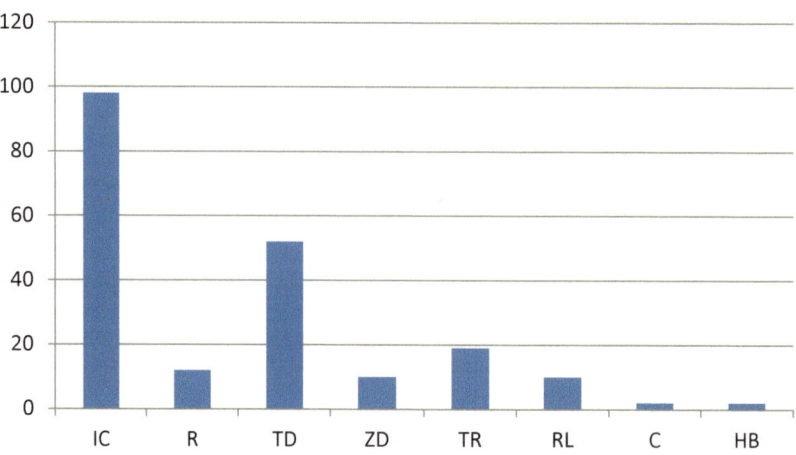

Fig. 7.9 Number of component failures of a metropolitan cellular network over three years

exposed areas. The relationship between temperature and copper density depends on the copper thickness, current is proportional to thickness raised to the power of 0.725 so roughly twice the current for twice the thickness would mean allowing twice the power without any significant improvement in the heat path out of the trace. An increase in temperature as well as the current capacity, such as having more copper in the PCB, more ventilation, or the inclusion of the heat path through the mounting hardware can have a negative effect of the remaining useful life (RUL). The current carrying estimates can be estimated from the IPC-2221 standard [1].

Corrosion is an important factor that affects the PCB's durability as it is apparent through visual inspection from the two sample PCBs shown in Fig. 7.11. A higher copper thickness allows a correspondingly higher current that flows through the copper tracks. Environments with high humidity and contact with acidic contaminants such as dust and sulphur from acid rain) can gradually etch away the surface of the copper tracks [44]. A surface finish such as electroless nickel plating can be applied during the PCB fabrication process in order to protect copper from corrosion.

Another observation from Fig. 7.11 is that dust particles can increase friction on contacting surfaces thereby accelerating surface wear and fretting corrosion, which in turn

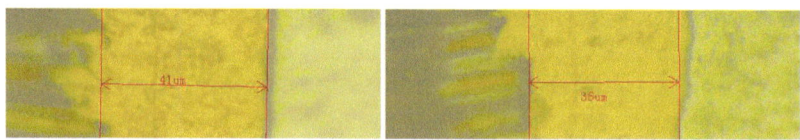

Fig. 7.10 Sample PCB cross-sectional copper thickness

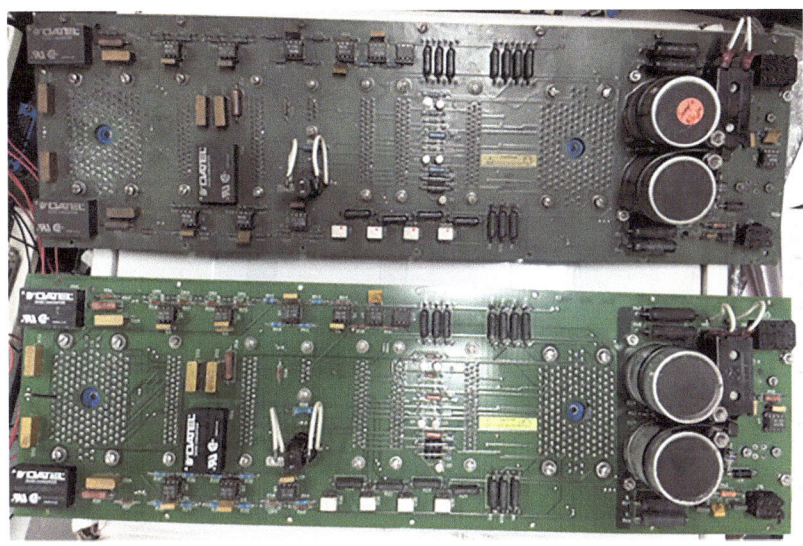

Fig. 7.11 A comparison of two identical PCBs subject to different use conditions

can increase the contact resistance [5]. Dust can induce electrical leakage, shorting and opening of PCBs under different operating conditions. A conductive electrolyte film can be accumulated that can degrade the surface insulation resistance between metal conductors. Certain types of dust particles can also act as dielectric materials that induce signal interference in the contaminated signal connectors and lines. Dust accumulation on the heat sink, power connectors, or active devices can cause overheating.

The uncontrollable operating environment with more dust exposure in various parts of an electrical circuit can cause severe impacts on electronics reliability within a communication device. Airborne particles as dust contamination that particulates in different components on the PCB needs to be addressed because of the mechanical operations and these are exacerbated by encapsulation under solder masks and conformal coatings [9]. Moisture can be a particularly important problem in tropical and coastal areas where the combination of heat and humidity can cause more component degradation.

Summary 7.4 6G Telemedicine Applications

1. Within any communication devices, circuits and systems, electronic component life cycle loads can be significantly affected by operating environmental conditions such as temperature, humidity, vibration, shock, utilization duration and frequency. The daily duty cycle of the components needs to be studied, the actual working conditions can vary substantially when installed in different locations.

2. Components such as IC, capacitor and relay are generally found to exhibit the highest failure rate among network equipment.
3. The aging profile of a given component can differ substantially in a controlled laboratory test verses that subject to actual field use. The actual operating environment must therefore be considered when determining component degradation. Field tests are therefore necessary to be conducted at all device, circuit and system levels.

7.5 Reliable Healthcare in Smart Cities Supported by 6G Networks and Systems: A Summary

The design and implementation of 6G networks and systems open up opportunities for providing reliable healthcare services in a smart city environment. 6G networks and systems are set to serve as the core backbone to integrate both existing and new infrastructures, as well as to link a variety of different sectors together. Services such as smart energy for more efficient network management [29], smart transportation that makes smart ambulance safer and efficient through optimizing V2I communications [17], in conjunction to secured smart health [7]. All these services and solutions form the basis of connected health in a smart city environment supported by 6G.

Central to the successful deployment of 6G networks and systems entail artificial intelligence (AI) solutions, IoT wireless sensing networks (WSNs), and operational reliability assurance. For example, an effective way of managing COPD (Chronic Obstructive Pulmonary Disease) in a smart city is to monitor the concentration of various toxic pollutants such as PM2.5, PM10, NO_2, and CO_2 using pollution sensing networks [43]. While 6G systems are fully capable of supporting such large array of sensing networks across the smart city, health management also requires AI algorithms the analyze spatial, temporal as well as geographical relationships from the vast amount of data acquired from each sensor within the network. Data acquisition with dynamic geo-fencing for the mapping of localized pollution patterns similar to the use case demonstrated in pathogen spread analysis [51]. Given that health related issues are closely linked to the release and spread of pollutants, one efficient way of controlling the discharge of toxic gases released from combustion engines is to integrate smart transportation to ease congestion and diversion of traffic to enhance both mobility efficiency and reduce the discharge of pollutants. Again, cellular communication networks form the basis of integrating different smart services across the entire smart city infrastructure.

The complex problem associated with environmentally-linked health issues is mainly caused from local sources such as vehicle emissions, power stations, industry, and civil activities, as well as regional pollutant migration from different industry sectors [41]. The primary goal of deploying AI algorithms analyzing data from the environmental sensing networks is to utilize the many and disparate air quality related data sources that are available to enhance monitoring, detection, and control of urban air quality, and in addition to improve decision making and managerial deployment of sustainable air quality strategies. This requires focusing on data mining and knowledge discovery with respect to air quality surveillance, as well as quantification of economic, social, and financial implications of air quality control. Algorithms are needed in extracting and identifying correlations and causal relationships among various social and environmental issues through sophisticated data mining and statistical learning methods. 6G networks are set to support surveillance systems for monitoring air quality across the entire smart city and coupled with suitable management strategies in quantifying social impact and economic risks caused by air quality and developing optimal policies and countermeasures for health and environment protection.

Integrating with smart transportation network, real-time traffic movement data, including GPS in conjunction with cellular communication, vehicle-to-vehicle (V2V) traffic conditions, and surveillance video images. A city-wide data warehouse system can be developed through integrating and managing real-time traffic movement data together with multiple streams of air quality data for the development of knowledge discovery, casual relationships, real-time surveillance, and policy improvements. In particular, statistical learning methods are needed to extract systematic correlations and relationships among air quality related variables, this can be accomplished through structural equation modeling (SEM) methods for identifying influential factors on quality characteristics [35].

One major challenge is to achieve a comprehensive view of air quality by combining time series of multiple pollutants at multiple sites, spatiotemporal measurements with different instruments, and different summary statistics of pollutant levels. To integrate multiple data streams and capitalize on knowledge of the complex dependence across time and space, it is of both theoretical and practical importance to develop a hierarchical model with conditional sub-models defined hierarchically at different levels. The uncertainty underlying the model is apportioned to different levels and propagated through the hierarchy, which provides a formal way to borrow strengths among various components and improve the statistical inference precision. This complex model structure poses a challenge for statistical inference. To overcome this problem, Markov Chain Monte Carlo (MCMC) algorithms can be used to estimate the model through of the conditional independence structure in the hierarchical model [40]. The overall system design and implementation approach is shown in Fig. 7.12. This system facilitates the visualization of machine learning-based pollutant dispersion analysis computed in conjunction with relevant weather information such as wind velocity, rate of rainfall and atmospheric air pressure to fine-tune the management model for air pollution analysis across the smart city region thereby provide important insights into combating various respiratory-linked diseases.

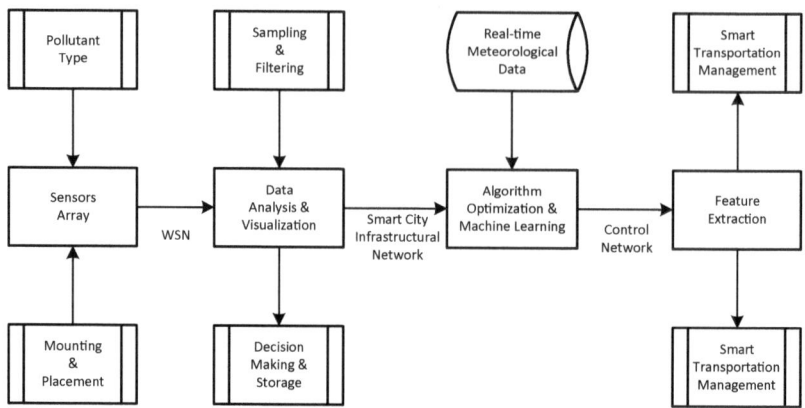

Fig. 7.12 Design and implementation of 6G pollution sensing system for respiratory disease management

Summary 7.5 Reliable Healthcare in Smart Cities Supported by 6G Networks and Systems: A Summary

1. 6G networks form the elementary building blocks of a smart city that links new systems into existing legacy infrastructure, thereby support integrated services for the enhancement health management for both residents and visitors. Central to the successful deployment of such 6G networks depend on the optimization of operational reliability, network availability and interoperability for harmonious connections between different types of devices and systems.
2. The performance of a given 6G network can be severely affected by a range of uncontrollable factors, this is particularly problematic when supporting mission-critical healthcare services. The study of network performance and reliability entails analysis of both long-term historical data as well as the current operating environment due to the fact that performance degradation and component aging can differ substantially even among identical devices, circuits and systems.

References

1. J. Adam, Thermal management of boards and current-carrying capacity of traces. Bodós Power Syst. (2011)
2. D. Ajitha, C.S. Gouri, S.B. Eklure, C. Chakraborty, Healthcare infrastructure in future smart cities, in *Intelligent Healthcare*. (Springer, 2022), pp. 321–341

3. E. Arndal et al., Chronic rhinosinusitis in COPD: a prevalent but unrecognized comorbidity impacting health related quality of life. Respir. Med. **171**, 106092 (2020)

4. R.N. Bharagava, G. Saxena, S.I. Mulla, *Introduction to industrial wastes containing organic and inorganic pollutants and bioremediation approaches for environmental management* (Springer, In Bioremediation of Industrial Waste for Environmental Safety, 2020)

5. K. Breivik, PCB in open applications: residential and public buildings. Sci. Total. Environ. **377**(2–3), 296–307 (2007)

6. P.M. Carter, C.A. Flannagan, C.R. Bingham, R.M. Cunningham, J.D. Rupp, Modeling the injury prevention impact of mandatory alcohol ignition interlock installation in all new US vehicles. Am. J. Public Health **105**(5), 1028–1035 (2015)

7. R. Chaudhary, A. Jindal, G.S. Aujla, N. Kumar, A.K. Das, N. Saxena, Lscsh: Lattice-based secure cryptosystem for smart healthcare in smart cities environment. IEEE Commun. Mag. **56**(4), 24–32 (2018)

8. T. Delamotte, M.G. Schraml, R.T. Schwarz, K.U. Storek, A. Knopp, Multi-antenna-enabled 6G satellite systems: roadmap, challenges and opportunities, in *WSA 2021; 25th International ITG Workshop on Smart Antennas* (2021), pp. 1–6

9. B. Ellis, Environmental issues in electronics manufacturing: a review. Circuit World (2000)

10. W. Fan, P. Kyosti, M. Rumney, X. Chen, G.F. Pedersen, Over-the-air radiated testing of millimeter-wave beam-steerable devices in a cost-effective measurement setup. IEEE Commun. Mag. **56**(7), 64–71 (2018)

11. Fluoridation Census (1985) US Department of Health and Human Services. Public Health Service

12. B. Fong, P.B. Rapajic, G.Y. Hong, A.C.M. Fong, Factors causing uncertainties in outdoor wireless wearable communications. IEEE Pervasive Comput. **2**(2), 16–19 (2003a)

13. B. Fong, P.B. Rapajic, G.Y. Hong, A.C.M. Fong, The effect of rain attenuation on orthogonally polarized LMDS systems in tropical rain regions. IEEE Antenn. Wirel. Propag. Lett. **2**, 66–67 (2003b)

14. B. Fong, P.B. Rapajic, G.Y. Hong, A.C.M. Fong, On performance of an equalization algorithm based on space and time diversity for wireless multimedia services to home users. IEEE Trans. Consum. Electron. **49**(3), 597–601 (2003c)

15. B. Fong, A.C.M. Fong, G.Y. Hong, H. Ryu, Measurement of attenuation and phase on 26-GHz wide-band point-to-multipoint signals under the influence of rain. IEEE Antenn. Wirel. Propag. Lett. **4**, 20–21 (2005a)

16. B. Fong, A.C.M. Fong, G.Y. Hong, On the performance of telemedicine system using 17-GHz orthogonally polarized microwave links under the influence of heavy rainfall. IEEE Trans. Inf. Technol. Biomed. **9**(3), 424–429 (2005b)

17. B. Fong, L. Situ, A.C. Fong, Smart technologies and vehicle-to-X (V2X) infrastructures for smart mobility cities, in *Smart Cities: Foundations, Principles, and Applications* (2017), pp. 181–208

18. B. Fong, A.C.M. Fong, C.K. Li, *Internet of things in smart ambulance and emergency medicine* (Technologies and applications, Wiley-IEEE Press, Internet of things A to Z, 2018), pp.475–506

19. B. Fong, A.C.M. Fong, K.F. Tsang, Capacity and link budget management for low-altitude telemedicine drone network design and implementation. IEEE Commun. Stand. Mag. **5**(4), 74–78 (2021)

20. B. Fong, H. Kim, A.C.M. Fong, G.Y. Hong, K.F. Tsang, Self-cognizant prognostics for the design and implementation of mission-critical telemedicine systems under the influence of heavy rainfall. IEEE Commun. Mag. **60**(8), 80–84 (2022)

21. T.K. Fong, P.S. Henry, K.K. Leung, X. Qiu, N.K. Shankaranarayanan, Radio resource allocation in fixed broadband wireless networks. IEEE Trans. Commun. **46**(6), 806–818 (1998)

22. T.R. Frieden, A framework for public health action: the health impact pyramid. Am. J. Public Health **100**(4), 590–595 (2010)
23. R.L. Gowland et al., Broken childhoods: rural and urban non-adult health during the Industrial Revolution in Northern England (eighteenth-nineteenth centuries). Bioarchaeology International. **2**(1), 44–62 (2018)
24. W. Guo et al., Molecular physical layer for 6G in wave-denied environments. IEEE Commun. Mag. **59**(5), 33–39 (2021)
25. F. Hawley, S. Palaszewski, Characterization of realistic environmental conditions for predicting the reliability of outdoor cellular electronics. Electron. Cool. 12–17 (2011)
26. Y. Huang, F. Ji, Z. Wei, M. Wen, W. Guo, Signal detection for molecular communication: model-based vs. data-driven methods. IEEE Commun. Mag. **59**(5), 47–53 (2021)
27. ITU, Propagation data and prediction methods for the planning of indoor radio communication systems and radio local area networks in the frequency range 300 MHz to 450 GHz. Recommendation ITU-R P.1238–11 (2021)
28. H. Kunsei, P.R. Hoole, K. Pirapaharan, S.R.H. Hoole, Tracking everyone and everything in smart cities with an ANN driven smart antenna, in *Machine Learning Techniques for Smart City Applications: Trends and Solutions* (Springer, 2022), pp. 75–94
29. C.S. Lai, L.L. Lai, Q.H. Lai, *Smart Grids and Big Data Analytics for Smart Cities* (Springer, 2021)
30. C. Liaskos, S. Nie, A. Tsioliaridou, A. Pitsillides, S. Ioannidis, I. Akyildiz, A new wireless communication paradigm through software-controlled metasurfaces. IEEE Commun. Mag. **56**(9), 162–169 (2018)
31. W. Li, H. Yin, Z. Qin, Y. Cao, M. Debbah, A multi-dimensional matrix pencil-based channel prediction method for massive MIMO with mobility. IEEE Trans Wirel Commun (2022)
32. J. Lin, H. Chaudhry, A. Gill, M. Roytman, D. Prajapati, D. Slater, A head scratcher: hepatocellular carcinoma manifesting as a subcutaneous scalp lesion. ACG Case Rep. J. **9**(8) (2022)
33. F.W. Lipfert, Air pollution and community health: a critical review and data sourcebook (1994)
34. Z. Lv, D. Chen, J. Li, Novel system design and implementation for the smart city vertical market. IEEE Commun. Mag. **59**(4), 126–131 (2021)
35. Z.A. Malaeb, J.K. Summers, B.H. Pugesek, Using structural equation modeling to investigate relationships among ecological variables. Environ. Ecol. Stat. **7**(1), 93–111 (2000)
36. N. Muhammad, Z. Fang, M. Shoaib, Remaining useful life (RUL) estimation of electronic solder joints in rugged environment under random vibration. Microelectron. Reliab. **107**, 113614 (2020)
37. W. Nam, D. Bai, J. Lee, I. Kang, Advanced interference management for 5G cellular networks. IEEE Commun. Mag. **52**(5), 52–60 (2014)
38. S. Naveau, G. Gaudé, A. Asnacios, H. Agostini, A. Abella, N. Barri-Ova, T. Poynard, Diagnostic and prognostic values of noninvasive biomarkers of fibrosis in patients with alcoholic liver disease. Hepatology **49**(1), 97–105 (2009)
39. J. Ostrometzky, R. Raich, L. Bao, J. Hansryd, H. Messer, The wet-antenna effect—A factor to be considered in future communication networks. IEEE Trans. Antenn. Propag. **66**(1), 315–322 (2017)
40. D. Öttl, R.A. Almbauer, P.J. Sturm, G. Pretterhofer, Dispersion modelling of air pollution caused by road traffic using a Markov Chain-Monte Carlo model. Stoch. Env. Res. Risk Assess. **17**(1), 58–75 (2003)
41. M. Popović, B. Rajović, Impact of the Industry 4.0 on smart city development. J. Engin. Manag. Competitiv. **11**(1), 64–76 (2021)
42. K. Rikkinen, P. Kyosti, M.E. Leinonen, M. Berg, A. Parssinen, THz radio communication: link budget analysis toward 6G. IEEE Commun. Mag. **58**(11), 22–27 (2020)

43. D. Sarabia-Jácome, A. Belsa, C.E. Palau, M. Esteve, Exploiting IoT data and smart city services for chronic obstructive pulmonary diseases risk factors monitoring, in *2018 IEEE International Conference on Cloud Engineering (IC2E)* (2018), pp. 351–356.
44. R. Schueller, Creep corrosion on lead-free printed circuit boards in high sulfur environments. SMTA News J. Surface Mount Technol. **21**(1), 21 (2008)
45. T.P. Sharkey, W.M. Simpson, Accidental sodium fluoride poisoning: report of eight cases, with one fatality. J. Am. Med. Assoc. **100**(2), 97–100 (1933)
46. E.C. Strinati et al., Reconfigurable, intelligent, and sustainable wireless environments for 6G smart connectivity. IEEE Commun. Mag. **59**(10), 99–105 (2021)
47. Y. Umeda, S. Takata, F. Kimura, T. Tomiyama, J.W. Sutherland, S. Kara, J.R. Duflou, Toward integrated product and process life cycle planning—An environmental perspective. CIRP Ann. **61**(2), 681–702 (2012)
48. T.J. Van der Kuijp, L. Huang, C.R. Cherry, Health hazards of China's lead-acid battery industry: a review of its market drivers, production processes, and health impacts. Environ. Health **12**(1), 1–10 (2013)
49. S. Vegunta, S.P. Lester, S. Pruthi, D.M. Mussallem, (2020) Effects of major lifestyle factors on breast cancer risk: impact of weight, nutrition, physical activity, alcohol and tobacco. *Breast Cancer Management, 9*(4), BMT51
50. Y. Wang, Q. Zhu, Modeling and analysis of small cells based on clustered stochastic geometry. IEEE Commun. Lett. **21**(3), 576–579 (2016)
51. H. Wang et al., A new COVID-19 quarantine directive: QDex evaluated dynamic geofencing. IEEE Commun. Mag. **60**(1), 94–99 (2022)
52. C.K. Wu et al., Supply chain of things: A connected solution to enhance supply chain productivity. IEEE Commun. Mag. **57**(8), 78–83 (2019)
53. B. Yang, M. Usman, Do industrialization, economic growth and globalization processes influence the ecological footprint and healthcare expenditures? Fresh insights based on the STIRPAT model for countries with the highest healthcare expenditures. Sustain. Produc. Consump. **28**, 893–910 (2021)
54. B. Zheng, C. Zeng, S. Li, G. Liao, The MMV tail null space property and DOA estimations by tail-ℓ_2, 1 minimization. Signal Process. **194**, 108450 (2022)